D0457219

MORE ADVANCE PRAISE FOR *WEATHER'S GREATEST MYSTERIES SOLVED!*

"Another superb accomplishment by this acclaimed professor and author . . . eclectic, entertaining, and enlightening."

—Russell S. Vose, chief, Climate Analysis Branch,
National Climatic Data Center

"Whether your interest is the past, present, or future, there is no escaping the weather. And despite how much we talk about it, who really understands the weather and how it impacts our lives? In this remarkable book—part detective story, part introduction to climate, part archaeological account of early civilizations—Randy Cerveny provides a rapid-paced overview of how weather has influenced human societies and may continue to do so in the future. Essential reading for anyone interested in global warming, this is science writing at its best."

—David S. Whitley, PhD, author of
Cave Paintings and the Human Spirit: The Origin of Creativity and Belief

"A fascinating collection of vignettes illustrating the dramatic impact that short-term 'natural' climate change has had on humanity. Reading these makes one reevaluate the risk associated with anthropogenic climate change."

—Dr. Joe Schaefer, past president,
National Weather Association

WEATHER'S
GREATEST MYSTERIES
SOLVED!

WEATHER'S GREATEST MYSTERIES SOLVED!

RANDY CERVENY

59 John Glenn Drive
Amherst, New York 14228–2119

Published 2009 by Prometheus Books

Inquiries should be addressed to
Prometheus Books
59 John Glenn Drive
Amherst, New York 14228–2119
VOICE: 716–691–0133, ext. 210
FAX: 716–691–0137
WWW.PROMETHEUSBOOKS.COM

13 12 11 10 09   5 4 3 2 1

Library of Congress Cataloging-in-Publication Data

Cerveny, Randall S.
Weather's greatest mysteries solved! / by Randy Cerveny.
p. cm.
Includes bibliographical references and index.
ISBN 978–1–59102–720–1 (hardcover : alk. paper)
1. Climatology—Miscellanea. I. Title.

QC870.C47 2009
304.2'5—dc22

2009004493

Printed in the United States of America on acid-free paper

To my fellow brilliant scientists who are making great strides in unraveling the most complex mystery of our time— the climate of planet Earth.

CONTENTS

Chapter 1

WHAT IS A WEATHER MYSTERY?

Weather and climate have an impact on us every day of our lives. How unusual was the hurricane season of 2005? Do you need to wear a raincoat to work today? Will there be a heat wave—or a cold snap—this coming weekend? What would a drought in California do to this winter's food prices? Will your friend's spring wedding be rained out? What kind of weather will your children or grandchildren experience? These questions point out the strong influence of weather and climate on the multitude of activities that are a part of our daily lives. But with such an all-encompassing presence in our lives, what can be mysterious about weather and climate?

Time: Spring 2009
Location: New York City, New York, North America

Nine-year-old Toby dejectedly stared out at the miserable, dark clouds and watched them with growing

dismay. Soaking rains continued to fall from the leaden skies. As he stared forlornly through the living room window, a growing cascade of wet rivulets spilled down the glass. In the background, he could faintly hear the cheery voice of a local newscaster on the TV.

"Drenching rain will continue all day today and Sunday so all of you with outdoor weekend plans will have to put them on hold for a while. It's going to be a wet weekend! Yes, definitely a good time to stay indoors by the fire with a good book. Traffic on the Long Island Expressway is particularly heavy as motorists negotiate over the water-filled roads—so let's all be careful out there! And now to Dan at the Sports Desk . . ."

"Thanks, Ed. And, as we reported earlier, that rain you just mentioned has forced the cancellation of today's big Yankees game with the Red Sox . . ."

Toby looked forlornly down at his brand-new baseball mitt, the one that he had hoped would catch the winning home run at today's game.

"Stupid old rain!" he muttered sadly.

A rumble of distant thunder laughed back at him.

After watching or reading the 24-7 reports from newspapers, television, Internet Web sites, or magazines, it would seem obvious that the natural environment around us has changed dramatically in the past few years. Coverage of hurricanes, floods, droughts, and the like has increased exponentially. There is always a weather-related natural disaster occurring somewhere in the world. The reporting of events like the horrible aftermath of Hurricane Katrina, the terrible European heat wave of 2003, and the massive flooding and tornadoes in Iowa and Georgia during spring 2008 suggests that we live in a world fundamentally different from that of our grandparents or even our parents.

Yet is that really the case? The extraordinary people who are interested in asking—and answering—that kind of question are known as climatologists, that is, scientists who study the long-term aspects of weather. But how do climatologists differ from the guys or gals who stand in front of weather maps each night on television? Is "climate" really that different from "weather"? A favorite author of mine, Robert Heinlein, once proposed that "climate is what we expect, weather is what we get."[1] These are unquestionably simple definitions, but they can be used as a foundation for our discussion.

At its core, climate may be defined as the "collections" or "groups" of different types of weather. Consequently, we can study climate through time such as "How is our weather changing over the past decade or century?" Or we can look at climate through study of a specific type of weather as, for example, "Are hurricanes becoming more intense or more frequent?" In short, "climate" is the general and "weather" is the specific. It is no accident that a substantial amount of the training needed to become a meteorologist—a weather scientist—is essentially the same training required for a career as a climatologist. Not surprisingly, a good climatologist is frequently also a good meteorologist.

Nevertheless, there are differences. In order to study the "general" of climate, we first need to study the "specific" of weather. Consequently, weather research—the study of the immediate atmosphere of today and the near future—is more advanced than the investigation of long-term weather patterns in our atmosphere, which is one of the central cores of climate research. We have been scientifically investigating weather slightly longer than we have been studying climate, perhaps by a hundred years or so. But both fields are young compared to other academic subjects.

This "newness" is real despite recent media attention on global warming and weather change that might give the impression that climatology and meteorology have been around for a long time. In fact, the vast majority of climatologists around the world have come from these other subjects, for example, atmospheric science, geographical science, physics, or geophysics. There aren't very many universities that currently offer degrees in climatology.

Each fall when I start teaching a new class of freshmen about weather and climate at my university, one of the things that I initially stress is the peculiar "newness" of the sciences of climatology and meteorology. Let me demonstrate that for you in a very straightforward, quick manner. First, pick up a basic college physics or chemistry textbook and flip it open to the first chapter. As you start reading, notice the time frame of those first primary progenitors of the field—the "fathers" of that particular science. What you will undoubtedly find in those chemistry or physics textbooks is mention of scientists who lived—and died—several hundred years ago. I'm referring to people such as Sir Isaac Newton or Galileo. Physics and chemistry—while without a doubt critically important—have thus been extensively studied for a considerable time. Given that long time frame, we know a great deal about those specific fields of study. (That's one reason why those college physics and chemistry textbooks are usually quite thick, as well).

Now pick up a climate textbook and open it to the first chapter. As you start reading of the people who are considered the mothers and fathers of the science, you will likely notice something odd, at least compared to physics or chemistry: most of the founders of climatology are still alive and well. The initial fathers and mothers of the science—brilliant people like Stanley Changnon, Yale Mintz, Ellen Mosley-Thompson, Stephen Schneider, Susan Solomon, Ann Henderson-Sellers, and others whom you would read about in those first chapters—are likely very much alive or may only have recently passed away. By the way, note that the field of climatology does in addition to fathers have mothers, who are not nearly as prevalent in other physical sciences like physics and chemistry. Thus, the issue at hand is that climatology is a young science.

Even its slightly older counterpart, meteorology, is scientifically "young" compared to many other sciences. Founders of modern meteorology such as Vilhelm Bjerknes, Carl-Gutstaf Rossby, L. F. Richardson, Bob Maddox, and Ted Fujita conducted most of their research within the last century. Many of the major advances in our knowledge of weather have only been made since the turn of the twentieth century.

That means that both meteorology and climatology are also still mysterious sciences, even to their own experts. By mysterious, I don't mean spooky—although sometimes climate and weather studies can involve very strange events or phenomena. Exploding fireballs, massive ice sheets, and towering "black blizzards" are just a few of the mysterious aspects of meteorology and climatology. But by mysterious, I simply mean we don't yet have a complete understanding of all the intricacies of weather and climate, and, as such, they are mysterious.

Mysteries are all around us. Some mysteries can be simply stated as straightforward questions. For example, "Who committed the crime?" or "Where is the missing clue?" are two of the favorite questions found in many mystery books, movies, and television shows. But many important mysteries of today don't need a crime to have been committed. Some of those critical mysteries of our world simply require a basic curiosity in what makes our environment tick. The solutions to these mysteries are based on gaining a fuller understanding of the world around us.

Climatologists and meteorologists have a passion for asking and answering basic questions about our environment. They become, in essence, climate detectives or weather detectives. Sometimes the questions that they ask are seemingly straightforward; such as, "What makes a cloud form?" Other questions can be much more complex, such as "How do changes in one distant part of the world's oceans cause changes in weather around the rest of the world?" All of these questions are mysteries—even today, since we don't know their complete answers. Again, as I am teaching that new class of first-year college students about climate, I point out that we are likely using a different climate textbook than the one from the previous year because we keep finding new answers and even sometimes discover that our old "answers" might not be complete. So our textbooks must be rewritten often.

Back in the 1980s, a group of scientists, for instance, ran an exhaustive set of computer simulations to find out what would happen if we detonated all of the nuclear warheads into the atmosphere. In essence, they were probing the mystery of "What would be the environmental result of an all-out nuclear war?"[2] Their study was given the nickname TAPPS, based on the first letters of the researchers' last names. Now, in those days, computer models of climate were in their very infancy. We basically didn't know back then (all of thirty years ago!) the complete intricacies of our environment. But that climate question of nuclear war was a very critical one for those last days of the cold war.

So in using their simple computer climate model, the TAPPS scientists who were investigating the environmental impact of nuclear war arrived at the sobering concept of "nuclear winter." Nuclear winter, their model's results suggested, would be the catastrophic environmental consequence of all-out nuclear war. The massive amount of radioactive dust and ash from a nuclear war would stay in our upper atmosphere for years, perhaps decades, shielding the earth from sunlight. The end result of that high-altitude dust shield, their model results indicated, would be a new—and long-term, perhaps even permanent—ice age.

Of course, those findings made national and worldwide news. Although everyone knew that, by its very nature, nuclear war would kill millions of people, it appeared from the TAPPS study that such killing was only a part of the tragedy—nuclear warfare would also lead to massive environmental change, catastrophically forcing us into a new ice age. This was compelling new ammunition for propeace activists and the results were used to promote a variety of social and political agendas.

Beyond the horrific social implications, the "means" of that nuclear war/climate study—rather than the results—was the critical issue for some scientists. Just how well did the computer model represent the myriad of processes that make up our environment? Dozens of climate scientists carefully studied all aspects of the computer scenarios in computer laboratories around the world. In particular, they were just coming to grips with the idea of "climate feedback"—that a change to our environment could cause "somethings" to happen, which somehow could lead to still other "somethings" that would either eventually amplify or diminish the original change. For example, if the world through some process became slightly warmer, that would lead to a worldwide change in the distribution of rain and snow—less snow would fall around the world. Because snow, as a white substance, reflects sunlight back into space, less snow in a warmer world would result in more sunlight. And that sunlight would instead "soak" into the earth rather than be reflected back into space. The extra heat received on the earth due to less snow would cause the world thus to become warmer than the original warming process would have accomplished alone. This amplified change in climate is called a *feedback*.

Follow-up scientists to the TAPPS study discovered that the early nuclear war/climate results weren't perfectly accurate because the original computer model for the nuclear war simulations lacked some of these critical feedback processes. Rather than nuclear war resulting in nuclear winter, subsequent scientists began to hypothesize that nuclear war might instead create "nuclear autumn"—a cooling, but not a catastrophic cooling, of the earth.[3] Of course, those results didn't make much news (it seems that finding fewer deaths is never as newsworthy as predicting more deaths), but it does point out the critical fact that we simply didn't—and still don't—understand everything about the earth's climatic system.

Consequently, in contrast to chemistry and physics, where major revisions of thought aren't quite as common anymore (or perhaps somewhat less popularized), the field of climatology appears to thrive on heated discussion, debate, and controversy—and controversy that is often focused squarely in the public eye! For instance, I don't see many *Wall Street Journal* editorials addressing pros and cons of current chemistry or astronomical theories. I have, however, seen many editorials in that paper over the past few years addressing the science of climate change.

Our interest in the environment around us has led us to try to discover what makes it tick. We may ask, "Has it always been the same?" Is climate only now changing due to humans—or has it always been changing? Have people in the past been able to adapt to those changes—or did they die out, surely a warning to us who are living today?

Solving "weather mysteries" of the past can be extremely valuable to all of humankind. A philosopher of the last century, George Santayana, once said, "Those who cannot remember the past are condemned to repeat it."[4] We might now say, "Those who forget the climates of the past are condemned to suffer from them again." Yet historians have notoriously been hesitant to link climate and history. A scholar of history, B. D. Shaw, even wrote, "Climate is one of those ever-present factors in the historical process that most historians prefer to ignore and yet cannot."[5] While there is an intriguing academic scandal of the early twentieth century that may have helped to foster this abandonment of linking climate and history (which I will touch on in the Mystery of the Vanished Harappans), I think in today's world, most people realize that our future comfort and survival on this planet are closely tied to our understanding—and eventual solution—of the mysteries of climate and weather.[6] And learning how past cultures have coped with (or succumbed to) environmental stresses is perhaps a critical key to our own society's future.

As an example of a weather mystery, let's look at what happened to little nine-year-old Toby in the story at the beginning of this chapter. At first, it may seem a simple case of bad luck—the youngster's misfortune in having his weekend outing to a baseball game unexpectedly rained out. But a climatologist might take a look at such an occurrence and dig deeper. She might check to see if there were more anecdotal accounts of weekend rains, and, if there were such accounts, she might then realize that a deeper mystery exists—the Mystery of the Weekend Rains. An inquisitive climatologist might ask, "Does it rain more on the weekends than during the workweek?"

Several years ago, Bob Balling, a colleague at Arizona State University (and an accomplished statistical climatologist), and I investigated that specific mystery.[7] At first glance, one might say, "Well that's a very trivial subject to study—why should my hard-earned tax dollars go to finding out whether or not it rains more on weekends? After all, there are people starving in other parts of the world!" A common enough sentiment in today's world, I'm afraid. But I think I can justify our study's importance in several ways. First, of course, hundreds of activities and events, such as Toby's baseball game, are dramatically affected by weather. Thus knowledge of whether or not it will rain can save the sponsors and organizers literally millions of dollars. Weather also can be instrumental in critical safety and health concerns; understanding, for example, when rains will occur can be, for individuals such as the sick or the elderly, a matter of life or death. And beyond that, such knowledge can indicate whether our activities can have a long-term impact on weather.

Why is that the case? In the case of weekend rainfall, if it does indeed rain more on weekends than weekdays, then it is likely that we ourselves are causing that weather. The reason would have to do with the underlying causes of various periodicities or cycles, including natural variations (such as the monthly cycle of the moon) and human-made cycles, such as the five-day workweek with a two-day weekend. Fundamentally, the weekly cycle, that is, the seven-day sequence of a week, doesn't exist in nature; it was created solely as an artifact of Western civilization. Therefore, if we discover weekly cycles in weather events such as rainfall—that it rains more on the weekends—we can then strongly suggest that people might be causing those events.

So we had our research question—our weather mystery—"Does it rain more on the weekends than on the weekdays?" and we had good justification for doing such a study. Our anecdotal evidence from stories like Toby's suggested that there might indeed be more rain on the weekends than on the weekdays. That gave us a foundation, a hypothesis, for our study. The next thing that my colleague and I did was to track down potential weather data that might allow us to test our idea. We wanted a data record that was as untainted as possible from potential human error. For example, in some cases, observations from manual weather stations are made by different people on a weekend than on a weekday—and a potential bias might exist. People might conceivably record their observations at different times of the day on weekends as opposed to weekdays, or even perhaps with different equipment. To avoid such errors, we found a very detailed daily oceanic precipitation record for just off the Atlantic coastline that was made by a set of orbiting TIROS-N satellites for seventeen years from 1979 to 1995—a long enough time period to see if there were any distinct cycles in the rainfall pattern.

If the rainfall occurred randomly from day to day, we would expect to see that each day of the week would contribute one-seventh (14.3 percent) of the weekly total rainfall. However, when we combined all of the satellite data along the East Coast of North America, we saw a very distinctive intra-weekly variability. The greatest precipitation consistently occurred on Saturdays (25.9 inches per year, 16.0 percent) while Mondays received only 21.2 inches per year (13.1 percent). Consequently, Saturday rainfall averaged 22 percent higher than Monday rainfall. But are those findings actually real? To answer that scientifically, we performed an exhaustive set of statistical tests ranging from simple correlation tests to involved cycle investigations to substantiate our initial findings. All of them confirmed that a pronounced difference existed between the rainfall amounts on weekdays and weekends.

As we continued to investigate the precipitation over the whole Atlantic Ocean, we came across a fascinating occurrence: a weekly cycle of rainfall literally moved across the entire ocean. That peak in precipitation moved from day to day (Saturday to Sunday to Monday and so on) as we studied areas progressively eastward. We discovered that the peak in weekly rainfall once again occurred on a Saturday on the far eastern side of the Atlantic near Great Britain.

That was intriguing and gave us a potential solution to this rainfall mystery. What if "something" was causing the extra rain along the eastern seacoast . . . and then was slowly moving across the ocean (still causing that extra rain) until by the following Saturday, "it" was located over Great Britain? Remember, we had started with the assumption that if we saw a weekly cycle in precipitation, it was likely something that people caused since the seven-day week cycle doesn't occur in nature. So if something was causing more rain during weekends on the East Coast and then moving across the ocean, what could it be?

The answer, we discovered, was likely our own pollution. Researchers have found high concentrations of human-made pollution over the North Atlantic Ocean and have traced back a substantial amount of these pollutants to the urbanized eastern seaboard. Weather models indicate that these pollutants are pushed eastward across the ocean by the normal westerly winds of the region. We learned that many cities demonstrate a strong weekly pollution cycle. Air pollution experts have even nicknamed this weekly cycle of pollution over cities the *Sunday effect*. What air pollution scientists told us is that major cities show high late-week pollution levels as opposed to that of early week. Because those air pollution experts believe that this is predominately an effect of automobile driving—auto pollution builds up over the week and reaches a maximum by Friday and Saturday and then is blown off the coast by Sunday—the change in transportation patterns from workweek to weekend produces the weekly Sunday effect.

So we see a weekly cycle in rainfall that moves across the Atlantic—and a corresponding cycle of air pollution that does the same thing. However, different events can have similarities without having to be related. Our next big key to solving the weekend rain mystery was to develop a theory—a set of ideas—explaining why and how pollution could influence precipitation.

A great advantage in science is that our work stands on the shoulders of geniuses who have already explored a number of these ideas. With regard to how pollution could influence rainfall, we turned to research by renowned climatologist Stanley Changnon of the University of Illinois. In the early 1970s, he undertook an extensive investigation of rainfall over and around cities and discovered that urban pollution acts to increase the amount of rain downwind of a city.[8] He and his colleagues suggested that this could be due to one of two physical processes. First the extra heat that a city emits (what is termed the *urban heat island effect*) can lead to increased air turbulence and enhanced precipitation; and sunlight absorption by the dust and pollution around the city can lead to heating the air and to eventual cloud and rain development. We therefore suggested that extensive *regional* pollution movement into the Atlantic Ocean might produce the rainfall modification by one or both of these mechanisms. Now, take notice that we didn't *prove* that pollution is causing those changes. Other researchers have, however, continued our initial work and have also identified a possible relationship between pollution and rain.[9] The more confirming studies that are published, the more likely our original speculations were correct. And our study led to even more general studies on the Sunday effect.[10]

The Mystery of the Weekend Rains shows how modern climatologists often act as weather detectives—first finding a mystery, then searching for clues and analyzing the evidence, and, finally, arriving at an explanation for the mystery. Throughout this book, I will present some of the more intriguing climate and weather mysteries that climatologists and meteorologists have studied over the past few decades. These mysteries are presented in a general historical sequence from the far past to the far future, covering an immense span of both time and space. They range from the catastrophic weather events that led to the extinction of the dinosaurs sixty-five million years ago; to the climate change that influenced early human culture in the Sahara; to the discovery of a new type of weather following a fatal plane crash in New York in the 1970s; and to an ambitious recent attempt to ascertain the future climate encompassing the next ten thousand years. Geographically, these mysteries will transport you across the globe to such exotic locations as a remote Indonesian volcano that almost wiped humanity off the earth; to southern Argentina, where Charles Darwin witnessed a type of weather that revolutionized our ideas about climate change; to London, England, where Benjamin Franklin tried to explain why American ships sailed so much faster than their British counterparts; and to even the mountainous region of Tibet, where ingeniously trained climate spies covertly monitored the weather of that remote corner of the planet.

In addition, most of the mysteries will have their own unique weather detective. These remarkable puzzles of weather and climate sometimes required unusual means of investigation and, consequently, unusual people to conduct those studies. In these mysteries you'll meet scientists who painstakingly stared through microscopes, carefully sorting minute bits of pollen for clues about past climates; climate scientists who delved into old history books for evidence of events that *didn't* occur; weather scientists who searched for—and eliminated—errors and biases in our past weather records; and even a geographer whose rather bizarre weather studies may have helped in the global war on drugs.

Finally, with each mystery, there usually will be a bit of technical information on a specific "climate tool," that is, the means of getting critical information. This is essential in shedding light on the overall climate mystery. The variety of "climate scene investigation" equipment that is used in these mysteries shows how varied and unique the life of a climatologist can be—with activities ranging from digging ice cores in Antarctica to coring trees in Arizona. As we work our way through history's weather mysteries, it is my hope that you'll begin to appreciate both the brilliance—and the limitations—of today's climate and weather sciences.

I need to give two final caveats before we begin. In these pages, I will be discussing the intricate and varied research of many brilliant colleagues and fellow scientists in climatology and meteorology. Please keep in mind that my analysis of their research is only my interpretation. Science isn't just about establishing facts, but is also about fitting those facts together in a manner to solve our weather mysteries. Occasionally reorganizing the facts can lead to substantially different conclusions than the researchers themselves may have intended. Therefore, my interpretations of their work should not be taken necessarily as the preferred explanations of the work of the great scientists with whom I have collaborated in the development of this book or whom I mention. Any errors, inaccuracies, or misconceptions should be attributed directly to me while credit for a deeper appreciation and understanding of the subject undoubtedly is a tribute to them.

Second, all of the vignettes at the beginning and ending of each mystery are, unless otherwise noted, the products of my imagination and should not necessarily be taken as representing actual people, places, or events as they occurred. A few of the stories are loosely based on an actual historical letter or diary entry, and such information is noted but the vignettes are fictional. They are designed to infuse each mystery with a bit more of the human element that is central to our understanding of how weather and climate influence us.

With those caveats out of the way, let's begin with our first mystery, one that occurred in a time long, long ago . . . but the place was disturbingly close.

Chapter 2

THE MYSTERY OF THE DEAD T-REX

Time: Sixty-five million years ago
Location: The site of present-day Denver, Colorado, North America

Impact minus six hours.

It probably started out like any other day in the Late Cretaceous period of geologic time—the end of the Age of the Dinosaurs. Warm, moist . . . a few billowing clouds hung low on the eastern horizon, perhaps the remains of last night's thunderstorm. A weak, salty breeze blew onshore from the

shallow sea that cut into North America's midsection like a dagger. A North American Tyrannosaurus Rex raised her head from her bloody meal of a Utah raptor, distracted by the rapid scurrying of a tiny rodent across the beach before her. After a moment's hesitation, she returned to her breakfast, her huge teeth ripping into the raptor carcass. If her eyesight had been a bit more acute, T. Rex might have spied the large comet trail that had become visible over the past few weeks. It was now so bright that it was easily visible even in the early morning sky. But the eyesight of the T. Rex was likely poor, so perhaps she didn't see that comet trail suddenly become a bright flaring light that grew steadily in intensity until the ball of fire outshone the sun.

Weather in T. Rex's day was probably much more consistent than what we experience today. It is likely that a weatherperson of sixty-five million years ago would have had a fairly easy life. Once she learned the phrase "continued warm and muggy with a likelihood of thunderstorms," she was set for life. Why? The best available evidence indicates that the world of T. Rex was both much hotter and much wetter than that of the present day. Scientists estimate global temperatures were perhaps 12°F warmer than those of today.[1] The basis of those estimates relies in part on the location of fossils dating back to those days. For example, the remains of Cretaceous-age crocodiles—today found in warmer, tropical climates—have been found north of the Arctic Circle. Likewise, warm water coral reefs likely extended a full 15° north of their present tropical locations.

The muggy part comes from the immense amount of what is present-day land that would have been covered by water. Because there were no glaciers on the planet, water that is currently locked up in the massive ice sheets of Greenland and Antarctica was residing in the world's oceans—and they were much, much higher than they are today. A massive shallow sea cut North America in half, stretching from modern-day Texas along the eastern side of the youthful Rockies Mountains all the way to Canada. One can still find fossils of sharks and clams in the center of the present-day landlocked states of Nebraska and Kansas. And those incredibly hot temperatures beating down on all that water did two things. First, the humidity—the mugginess—would have been much higher in T. Rex's day and, second, all that heat and air would have likely led to some spectacular thunderstorms!

Impact minus two hours.

Ignorant of the impending disaster, T. Rex contentedly finished her breakfast and began a leisurely search for more food. As she wandered along the beach, she sniffed the air. The humid sea air was filling rapidly with buoyant cumulus clouds, probably precursors to a late afternoon thunderstorm.

Climate models—like the one I discussed earlier that simulated the climate after an all-out nuclear war—are nothing more (or less) than a set of very complex mathematical equations. Consequently, and perhaps not surprisingly, climatologists need to have at least four semesters of calculus under their belts to do their trade properly. These math equations describe what we know about how the winds blow, how storms move, and why rain falls. Climate models aren't perfect—given that we don't have a perfect understanding of climate—but they provide some of the best means available to delve into situations and times that we can't directly study. So climate model simulations of the age of the dinosaurs or of a greenhouse gas–saturated world are our best stand-ins for actually being there and measuring the temperature with a thermometer. The scientists at Pennsylvania State University are currently some of the best researchers in the world at understanding the weather of T. Rex's world.[2]

Using their results from a very detailed simulation of the Late Cretaceous period by a *general circulation model* (or GCM)—a complex set of equations that describe how our earth's environment operates—a few of my students and I determined the Cretaceous values of a collection of severe weather indices currently used in modern weather forecasting.[3] We computed that the atmosphere of T. Rex was likely much more unstable, and therefore more likely to produce severe thunderstorms—with damaging winds, rain, and lightning—than our present-day storms. T. Rex's world would likely have been a storm chaser's dream! But that world was about to end with a very loud bang.

Impact!

A brilliant fireball of light leapt across the southeastern sky, startling a flock of small birdlike dinosaurs that T. Rex had been carefully stalking. At the sudden movement, T. Rex raised her massive head and bellowed an answering challenge to the visitor from the heavens.

The six-mile-wide asteroid—imagine a rock larger than Mount Everest—slammed into the Gulf of Mexico, off the present-day coast of the Yucatan Peninsula.[4] As it did so, a backshock (like the recoil of a rifle) jolted through the asteroid, instantly warping its probable rocky, oval shape and ripping its comet tail of debris to shreds. The rock itself likely melted a huge hole through the ocean floor as it plummeted at incredible speed through the ocean waters. As it did, most of the asteroid vaporized. Within seconds, the impact crater reached its maximum depth of nearly thirty miles deep. Then the center of the crater raced upward—like the liquid that leaps back into the air when you splash milk into a glass. Enormous landslides cascaded off the rapidly rising center, which soon grew so high that it collapsed under its weight, creating a ringlike set of ridges resembling a target's bull's eye.

Impact plus one minute.

The brilliant flash, startling the huge dinosaur just a brief moment before, gave way to a monstrous earthquake that rumbled beneath the Colorado landscape. T. Rex stumbled over the raptor's carcass, surprised but not unduly alarmed. Many times in the past, the earth had shaken under her feet. Perhaps she thought that the mother of all T. Rexes was walking nearby.

Interestingly, a direct earthquake from the asteroid itself would have likely been one of the later occurrences in the sequence of events of an asteroid impact with the earth. This is because the shock waves associated with a distant earthquake are dampened as they travel through the earth. So what did our T. Rex feel? It is likely any large asteroid would have had several chunks of rock or ice that accompanied it—much like the Shoemaker-Levy comet strike of Jupiter in 1994. So prior to the main impact, it is possible that smaller, earlier strikes could have caused earthquakes that rattled the prehistoric earth for several days prior to the main impact.

Current scientific thought suggests that the impact of a six-mile-wide asteroid would have generated about one billion times the energy released by the atom bomb that destroyed Nagasaki at the end of World War II. The gigantic sledgehammer force of the impact must have crushed rock to the point of destroying even the crystal structure of the minerals, creating what scientists call "shocked quartz."[5]

Passage of a large six-mile-diameter meteorite would have generated about 10^{30} (that's 1 followed by 30 zeros) ergs of energy. For comparison, the total sunshine received by Earth in a day is about 1029 ergs, the detonation of a single thermodynamic weapon is on the order of 10^{24} ergs (or 1 million times smaller), the daily output of Hoover Dam is 10^{21} ergs (1 billion times smaller). This was without a doubt a *gigantic* explosion!

Impact plus four minutes.

T. Rex glanced up toward the southeast as the ground continued to shake. Even her poor eyes now noticed a huge bright orange light along the southeastern horizon with a brilliant white tail stretching into the sky with alarming speed. The orange light grew brighter and closer.

The asteroid's passage bored a tunnel through the atmosphere extending from the impact crater back into space. This created a near vacuum that sucked in extremely hot, low-density plasma from the impact site, as well as the surrounding air, and jetted the combination out into space. The rush of air into the plasma tube at velocities beyond the speed of sound created a huge sonic boom.

The speed of the burning rock, minerals, and gases ejected from the impact crater would have been incredible; current research suggests that the first effects of the asteroid's impact off the Yucatan took only ten minutes to reach present-day New Jersey. The atmosphere of North America was traumatized—shock waves and burning debris filled the air.

Some of the impact's physical remains were spread around the world—and those remains proved to be the smoking gun that established a massive impact had occurred. In 1980, L. W. Alvarez (who had earlier won a Nobel Prize) and colleagues published a very famous scientific paper that proposed the abrupt transition from the age of the dinosaurs to the age of the mammals (called the *K-T boundary*, K for the Germanic form of Cretaceous and T for Tertiary) was caused by the impact of a six-mile-wide chunk of rock or ice. Their evidence was based on the discovery of exceedingly high concentrations of iridium, an element found in the geologic layer of the earth that corresponds to the K-T boundary time frame. Such high amounts of iridium, they theorized, could only be associated with material of extraterrestrial origin.[6]

Impact plus six minutes.

T. Rex released a startled bellow. A thunderous roar filled the air as though the earth itself was dying. T. Rex pawed the ground nervously knowing that something was terribly different about this day. Then without warning, an unimaginable hurricane blast of wind knocked her down while ripping everything—huge trees, boulders, rocks, and her fellow dinosaurs—off the ground like matchsticks.

The shock wave by an object such as an asteroid hitting the atmosphere at even a slow velocity of twelve miles per second would have had a devastating effect on forests and large animals over a very large area. The K-T boundary event may have had speeds many times that velocity. The shock wave created by the asteroid smashing into the earth at a conservatively estimated speed of over forty thousand miles per hour would have smashed everything across perhaps an entire hemisphere. Indeed, pollen records in parts of western North America and Japan indicate that large plants were completely exterminated from the landscape for a very long period after the impact.[7] Apparently almost everything standing, even inches above the ground, was smashed flat and likely killed.

Impact plus seven minutes.

Bruised, battered, and near death, the injured T. Rex had a final glimpse of the horizon-to-horizon wall of fire. The very atmosphere itself seemed set aflame. It roared up from the southeast before the dinosaur was consumed by a global conflagration of fire.

The fireball remains of the asteroid, together with falling chunks of burning ejecta from the impact crater, would have ignited and blasted plants, animals, and surface fuels over the entire Western Hemisphere and perhaps even large parts of the rest of the world. The result would have been gigantic wildfires. Geologists, particularly in North America, can easily see the layer of sediment associated with this event, the K-T boundary, contains charcoal—a key geologic indicator of the incredible firestorm that raged over the earth at that time.[8]

In some areas, the fires could have burned for weeks, filled the air with smoke and gas, and temporarily blotted out the sun. The fires—and the impact itself—would have caused the atmospheric carbon monoxide levels to skyrocket perhaps to even toxic levels, at least for a short time. Any rain that fell would have been heavily acidic.[9]

High levels of poisonous nitrous oxides would have also stripped the leaves off of plants and trees and it might have led to asphyxiation of animals; animals might literally have drowned in an atmosphere of toxic acid. The nitric oxides might have scavenged ozone from the stratosphere, permitting potentially lethal amounts of ultraviolet radiation to reach the surface. In addition, if there were high levels of trace elements such as nickel in the asteroid, dispersal of those elements into the lakes, rivers, and even the oceans may have reached toxic levels.

Impact plus two hours.

Even in death, the doomed T. Rex would have no peace. As the dinosaur's charred bones lay smoldering in the aftermath of the fireball inferno, another line—this one colored a deep ocean blue tipped with white—appeared on the southeastern horizon, rapidly growing larger with each passing second. As the blue line neared the death site of T. Rex, any lingering survivors of the shock waves and inferno would have gaped incomprehensibly at what it represented: a mile-tall wall of water was about to cascade over the burning rubble of the ancient Colorado coastline.

Tsunami! That huge water wave that some miscall a "tidal wave"—although no tides are involved with an earthquake- or asteroid-generated wave—was the next disaster to enfold. The asteroid's impact on the ocean likely would have created one of the largest tsunamis the planet has ever experienced. This ring of watery destruction would have rushed outward from the impact site at perhaps more than six hundred miles per hour, growing into a monstrous wall of water as it approached land. That colossal wave would have smashed into the coastal regions of North and South America, stripping away anything still standing.[10] It also would have sucked up massive amounts of submarine sediments, which proceeded to cover the 120-mile-wide impact crater—hiding it from view until scientists finally identified it from gravity anomaly analysis in the mid-1980s.[11]

The Krakatoa volcanic eruption of 1883 produced a tsunami estimated at perhaps 135 feet tall when it hit Java and Sumatra. One sea captain reported that the 1883 tsunami as "like a mountain. The monstrous wave precipitated its journey towards the land. And before our eyes this terrifying upheaval of the sea . . . consumed in one instant the ruin of the town . . . all was finished. There where a few moments ago stood the town of Telok Betong was nothing but the open sea."[12]

The tsunami produced by the impact of the six-mile-wide asteroid may have been ten times higher than the Krakatoa wave!

Impact plus five days.

Safe from the tsunami in the highlands, lying west of the final remains of the dead T. Rex, a small mouselike rodent—a cynodont—scurried around the charred and broken landscape, stopping occasionally to nibble at the scorched meat of a larger, less fortunate animal. The small mammal had survived the terrible impact, the shock waves, and the inferno, safe and snug in her tiny burrow—one of the very few creatures on the planet to do so. But now hunger was forcing her out to hunt for food. Although she was extremely uncomfortable in the mind-numbing heat, she had survived and here was food—with luck she would continue to survive.

The fireball is thought to have raised global air temperatures by at least 50°F. Many places close to the eruption would have been substantially hotter than even that. This temperature increase came from two sources: first, the extreme amount of heat energy that the asteroid impact had dissipated into the atmosphere; and, second, the reentry of burning rock and ejecta that the impact had thrown into the air. Some estimates suggest that the heat from these two sources may have been as much as 50 to 150 times the present amount of solar energy received by Earth.

And while there is still debate about its duration, some scientists believe this terrible heat wave may have lasted for weeks. This temperature increase—in addition to the tsunamis and firestorms—probably was a massive killer, producing environmental conditions intolerable to many species that were adapted to only narrow temperature ranges. This would be particularly true for cold-blooded animals and large vertebrates—such as dinosaurs. Such a dramatic and large heat blast, combined with the disruptions in the food chain, would have led to catastrophic death tolls.

The heat would have also radically altered Earth's atmospheric gases.[13] The furnacelike atmosphere could have even caused basic chemical reactions. Atmospheric oxygen, nitrogen, and water vapor, for example, may have combined to form toxic nitric acid. Some extremely toxic rain may have fallen for a month or two after the impact. The amount of acid rain would have depended on the composition of the asteroid. If it was ice rich, global pH values might have plummeted from 0 to –1.5—similar to the values of pure hydrochloric or sulfuric acid. Perhaps a conceptual analogue to the atmosphere of the smashed earth of that time would be the present-day air of Venus—a hostile sky of dark clouds, acidic rains, and toxic air.

Indeed, most of the few survivors of the Earth likely lived through the disaster by being underground where the temperatures were more moderate and the air was less toxic.

Impact plus two months.

The small rodentlike creature stuck her head out of her burrow and twitched her whiskers at the changes she saw. This day, as those that had preceded it for the last few months, was dark—a dry and cold dark untouched by either snow or sleet. Many of the ferns that the rodent had been forced to survive on for the

past several weeks were now withering into brown death. Each day now, the rodent was being forced to extend her search for food—and water was also becoming increasingly harder to find.

The great quantity of dust shot into the stratosphere by the asteroid's impact would soon have encircled the world and blocked out a major portion of the sunlight over a large part of the earth. This brought an abrupt end to the intense heat and replaced it with a period of bone-chilling cold. Photosynthesis would have been nearly impossible for many plant species.

Meanwhile, after the acid rain ended, global precipitation would likely have been completely shut down according to computer simulations of such an impact event.[14] The ability for rain is linked to atmospheric stability—the process in which heat rises in our atmosphere. A fundamental principle in meteorology is that hot air rises. If the surface becomes cold due to the shielding but the air above it is warmer (because of the massive dust loading), then heat is going to flow downward—and rain clouds won't form.

Both the global cooling and the global drought may have contributed to the massive planetary extinction—perhaps as many as 90 percent of all living species died out. Even large marine dinosaurs such as the mossaur and plesiosaur died because of breaks in their food chain and the deadly toxicity of the ocean waters.

Impact plus one year.

The little rodentlike mammal had survived the shock waves, the firestorm, the tsunami, the global drought, and the cold. Now, a year after that fateful day, the earth was slowly shifting back toward something close to normality. The cynodont felt warm—perhaps if her small brain could have reasoned, it would have compared the current hot temperatures to conditions prior to the global disaster. And, finally, as the skies had slowly cleared, clouds built up during the afternoon and brought wonderful soaking rains. She glanced around at her small offspring scurrying around the area. Perhaps life would begin again.

Recent computer models—following the earlier discussion of nuclear war—suggest that the cooling caused by the dust cloud would not have been as severe or prolonged as models had projected back in the early 1980s. This nuclear autumn likely would have lasted for only months rather than years.[15] Then, as the dust would have slowly settled out of the sky, sunlight once again would have begun to warm the surface. Eventually, the earth would have turned even warmer than before the asteroid strike. This would be because the global fires that followed the impact would have added tons of carbon dioxide to the atmosphere, amplifying the natural greenhouse effect.

Over the next sixty-four million years the ghastly scars of that terrible day were slowly covered up or hidden as the continents continued to drift across the globe to their current positions. Gradually, the climate began to cool, until about two million years ago, when great ice sheets began to form and dissipate in a regular cycle over much of North America and northern Europe, further camouflaging the planetary wounds. Only recently have scientists managed to piece together the geologic and climatologic clues to unveil the mystery of that extraterrestrial encounter. Meanwhile, the small mammalian survivors of the planetary holocaust slowly flourished and evolved over that incredibly long time of sixty-four million years—gradually diversifying in kind and size to fill in the various biological niches left vacant by the massive extinction event.

And some of the survivors adapted slowly over time to become our ancestors, which leads us to our next mystery.

Chapter 3
THE MYSTERY OF HUMANITY'S "NEAR EXTINCTION"

As the current masters of this planet, we like to think that humans have acquired a knack for survival. The human species, after all, is technically known as *Homo sapiens*, or the "wise people." We have been around for several tens of thousands of years without committing an ultimate fatal error. Granted, during that time, we have suffered through countless wars, famines, pestilence, and misery—but the population as a whole has managed to overcome those severe challenges and to survive. It would seem that our track record for survival is pretty good.

Yet anthropologists have recently uncovered evidence that at one very specific time in our past, we *as a species* came disturbingly close to the complete and total extinction of humanity. At that one instance in history, the entire human population may have been reduced to perhaps only a couple of thousand individuals and those people may have been found in only one tiny part of the planet. Let's examine the Mystery of Humanity's "Near Extinction" and discover how an incredible climate change nearly caused the destruction of the entire human race.

Time: Approximately seventy-three thousand years ago
Location: Region of present-day Iraq, Asia

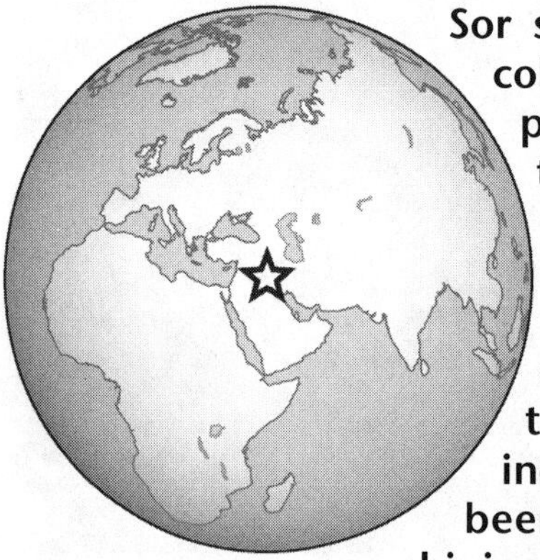

Sor shivered at the increasing bitter cold assailing her aching body and pulled her all-too-thin animal skins tighter about her shoulders. A freezing cold wind was continuing to blow from the frigid north. Yes, it would be terribly cold again tonight. She turned to the west, gazing across the increasingly barren plains. As had been case for the last days, the shining orb in the sky was sinking into the western ground in an angry state—no longer a bright yellow but instead a sullen reddish ball of flame. Why? Sor felt sad and confused. What had her people done? Why wouldn't the great ball of fire warm the land anymore?

Following the ear-piercing thunder that had shook the whole world several summers ago, a strange and seemingly unending thick dry rain of dirt coated the land. After that, a terrible cold settled on the land as if the earth itself decided to turn to ice. Her people endured—sheltered in their small cave—as the bitter cold persisted through one moon cycle after another.

Every day, the hunt for food was becoming harder and harder. Animals, which the whole tribe desperately needed to survive, were rapidly disappearing from the land. Sadly, under the continued onslaught of the cold, the tribe was uncovering dead animal carcasses more often than live—eatable—animals. Nuts and berries, even the very plants and grasses themselves, were shriveling up and dying under the assault of the persistent cold and the unwarming red sun.

We have upset the Great Earth,* Sor suddenly realized. *The Land no longer favors us.

Sor's tribe was in desperate straits. Telo, her mate, had died just a few days ago, finally succumbing to the awful coughing illness that had already killed most of their tribe. Oddly, after his death, Sor had felt compelled to place some flowers onto Telo's body at the back of the cave. Though flowers were hard to find, this gesture had just seemed a way to remember the good times when the earth was kind, although the few remaining members of the tribe responded with a few hoots of derision. But she realized even if the rest of the tribe did not, that a few cold and crusty flowers weren't enough food to save anyone in these punishing times. Even now, Sor noticed uneasily, her own chest felt heavy, constantly filling with a vile taste. She was coughing more and more frequently.

The cold red ball of fire disappeared beneath the freezing western lands in an unappreciated brilliant panorama of vivid purple and red hues.

Most scientists believe modern humanity developed from a base population that developed in eastern Africa perhaps as long as a million years ago. Then it slowly dispersed downstream along the game-rich Nile River valley, eventually migrating across the Sinai Peninsula into the Middle East, and in time into Asia and, perhaps, even to Australia. As humanity expanded into these regions, the populations in these distinct regions would over long periods of time start to display genetic differences from each other and from the original base population in Africa. This is because the population of each group would, by necessity, be limited in its genetic pool to people within a short geographic distance. Over a very long period of time, this space limitation would force distinct genetic differences to become apparent in the disperse populations.

One of spectacular scientific milestones in recent times has been the complete mapping of the human genome, the twenty thousand to twenty-five thousand genes present in the twenty-three chromosome pairs of all members of the human race. That mapping has led to intriguing findings regarding humanity's genetic roots. Despite humanity's extensive early geographic variability from Africa to as far as Australia, geneticists have found that we modern humans exhibit surprisingly little genetic diversity—particularly given the long period of time and space in which the human race has existed. For example, a group of scientists publishing in the highly respected journal *Science* concluded that all modern humans are still essentially African, at least with respect to their genetic DNA.[1] These geneticists stated that the African gene pool contains more variation than elsewhere, and the genetic variation found outside of Africa represents only a small subset of that which is found there. So one may conclude from such findings that Something (with a capital "S") occurred long after the original dispersal of peoples across the world to cause the abrupt loss of their genetic signature in our present DNA.

Recently, an anthropologist from the University of Illinois at Urbana-Champaign proposed a fascinating theory to account for the surprising amount of African genetic heritage found in all human DNA. Stanley H. Ambrose expanded upon that earlier genetic research as well as that of others, offering a fascinating climate-based theory to explain our pronounced lack of genetic diversity in the human population.[2] He first reiterated—and most anthropologists agree—that modern humans developed from *Homo erectus* nearly a million or so years ago, who subsequently dispersed from Africa to Asia and Australia. But then he said that about seventy thousand years ago—incredibly recently in geologic and evolutionary terms—a great calamity occurred that abruptly decimated Earth's human population, leaving perhaps only a few hundred individuals in Africa. Ambrose stated that it is to that second very small pool of individuals that we modern humans today owe our entire genetic heritage. Following that calamity approximately seventy thousand years ago, the small pool of remaining humans began the climb again to dominance and dispersion around the world. The proposed calamity that came so close to exterminating our entire species—yet simply called a *bottleneck* by anthropologists—caused many genetic lines of our species to abruptly end, leaving only a small number remaining. Ambrose stated that this bottleneck is the cause for our modern relative lack of genetic diversity.

But what could have caused such a horrific calamity? It would have been something that created a profound and major signature throughout nature around the world—and, therefore, an event that

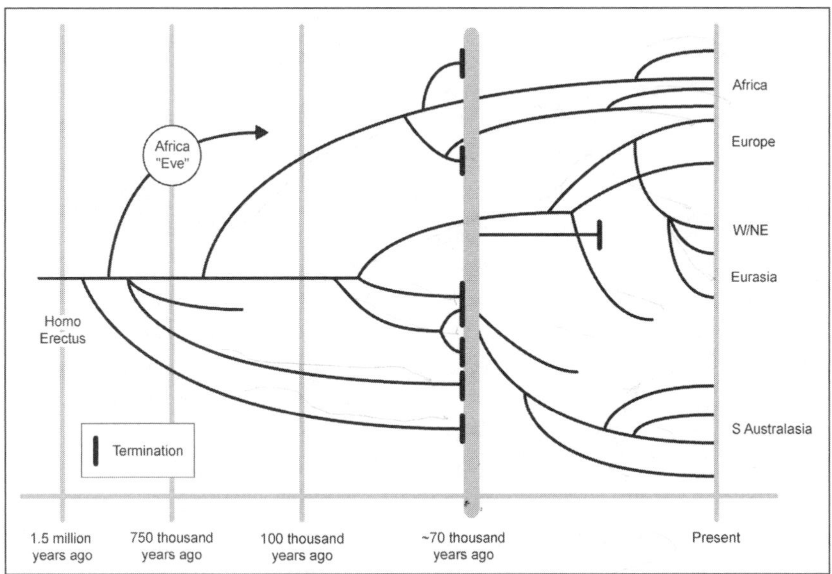

Time graph of various genetic lines of humanity from 1.5 million years ago to the present by geographic region with the bottleneck of seventy thousand years ago and various terminations of genetic lines noted. Graphic by Becky Eden.

researchers in other academic fields might have already identified. Ambrose knew the approximate date of the bottleneck—around seventy thousand to seventy-five thousand years ago. Do we know of anything that happened around that time that would literally bring the human race to within a hairsbreadth of extinction?

In today's world, we have discovered that climate change can occur in two distinct fashions. It can happen gradually over a very long period of time. For example, the so-called Milankovitch orbital mechanisms—the gradual change in the three basic characteristics of Earth's orbit (detailed more fully in chapter 21 discussing the Mystery of the Future)—have been documented as having a profound, but generally slow, impact on the earth's climate over the past two million years or so.

Conversely, we also know that occasionally something dramatic can happen to the earth almost in the blink of an eye, leading to monumental climate change—such as we saw in the Mystery of the

Dead T. Rex. In our current mystery, however, we don't think that climate change occurred as a result of a comet or asteroid impact. In this case, the perpetrator of mass murder—almost genocide—across the globe was a very large volcano named Toba, located near Sumatra in Indonesia.

Toba isn't a "normal" volcano like Mt. St. Helens, which erupted in 1980, or even the larger Mt. Pinatubo, which erupted in 1991. Instead, Toba is what has been called a "supervolcano."[3] When Toba erupted around seventy-three thousand years ago, scientists believe that it produced the largest eruption that the earth had experienced in the last two million years.[4] For example, volcanic ash from Mt. Toba can be found across India deposited in incredibly thick layers that are commonly three to ten feet deep but ranging, in some cases, to as much as eighteen feet thick.

To put Toba's eruption into perspective, the enormous eruption of a volcano called Tambora in 1815, which led to the infamous "Year without a Summer" of 1816 and killed thousands of people, threw nearly five cubic miles of ash into the air. That explosion has been the largest-known volcanic eruption in recorded human history. In comparison, Mt. Toba ejected nearly two hundred cubic miles into the air! It was forty times larger than the largest eruption we have seen in the last two centuries. The central crater—or more technically, the caldera—of the Toba volcano is eighteen miles by sixty miles. Compare that to the caldera of Mt. St. Helens, which is only two square miles. Size, however, isn't the only volcanic indicator we need to examine.

Four factors affect the climatic impact of a volcanic eruption. One is simply the amount of matter ejected by the volcano. Undoubtedly, Toba's eruption sent immense amounts of material into the atmosphere. But in addition to that amount, we must also account for the specific type of ejected material, the height to which the material was propelled, and how long the eruption lasted.

A volcano emits a wide range of materials and gases. The most prominent of these gases are water vapor, carbon dioxide, and sulfur dioxide. Some volcanic gases are harmless to humans, such as water vapor. Others have varying degrees of toxicity to humans. Nonetheless, the substance most influential to climate is sulfur. The chemical reaction is rather straightforward: sulfur dioxide combines with water vapor to form sulfate particles.

Those particles have the ability to reflect sunlight. So, from a climate standpoint, if a volcano was to inject a lot of sulfur into the air, the weather would likely get colder. Exactly how much colder depends in part on how much is put into the air. Two respected volcanologists, Michael Rampino and Stephen Self, studied three historical explosive eruptions of different sizes in Indonesia—the 1815 eruption of Tambora, the 1883 eruption of Krakatoa, and the more recent 1963 eruption of the volcano Agung.[5] Although the amount of overall material ejected into the high atmosphere differed greatly among the three volcanoes, all three resulted in roughly the same amount of global cooling (0.1–1.2°C, or somewhat less than 2°F). Rampino and Self discovered the similar cooling was the result of the specific amount of sulfate particles—not the total amount of volcanic ash, which tended to settle or rain out of the atmosphere within a few months of an eruption.

A comparison of two volcanoes of roughly similar magnitudes effectively demonstrates this finding. When the volcano El Chichon erupted in Mexico in 1982, it exploded at least eight megatons of sulfur particles into the stratosphere. That sulfur eruption resulted in a measurable cooling of the globe. In contrast, the similar-sized explosion of Mt. St. Helens in 1980, however, ejected only about a megaton of sulfur aerosols into the stratosphere. And, of great significance, there was not any significant cooling of Earth's surface after Mt. St. Helens erupted.

The third component of volcanic eruptions, which is important to climate change after the volcanic type and amount of material, is the height to which that material is propelled. Our atmosphere can be divided into different layers, depending on the atmosphere's temperature characteristics (basically, whether temperatures are increasing or decreasing with height). The four basic layers of our atmosphere are the troposphere (in which we live and throughout which temperatures decrease with height), the stratosphere (which extends from roughly ten to thirty miles and is a region where temperatures increase with height from –60°C, or –76°F, at ten miles to 0°C, or 32°F, at thirty miles), the mesosphere (from thirty to fifty miles up where temperatures decrease with height until, at fifty miles, the temperature is –90°C, or –130°F), and then the thermosphere (where theoretically temperatures are very hot as measured by molecular movement). Generally, air (and debris) has a hard time moving from one layer to another—particularly between the troposphere and stratosphere. If material gets sent up to the stratosphere, it can stay there for a long time—potentially for two or three years.

Consequently, if a volcanic eruption is so explosive that sulfur particles can be injected into the stratosphere, the resulting climate cooling near the surface can be great and long-lasting. For example, the large explosive eruption of Mt. Pinatubo in mid-June 1991 forced seventeen million tons of sulfur dioxide into the stratosphere. All of that stratospheric sulfur resulted in a global cooling of 0.5–0.6°C (about 1°F) that lasted for several years after the eruption.

Finally, the remaining critical issue for volcanic-induced climate change is how long did the volcanic activity last? Most modern explosive volcanoes have only erupted for relatively short periods (minutes or hours). Obviously, the longer the volcano is ejecting material into the stratosphere, the more intense and long-lasting the resulting cold period will likely be.

Therefore, to use the Toba volcanic activity as a possible explanation for the near extinction of humanity, we need to know several things: First, is there evidence that Toba actually "supererupted" seventy-three thousand years ago? Then, second, if it did erupt, how big was the eruption, what types of material were ejected, and how long did the eruption last? Finally, what was the climatic impact of such an eruption?

Of course, to be good climate detectives, we need to carefully select our investigative tool. In this case, we need something that will measure climate over very long time periods—literally tens of thousands of years—and, perhaps, even show us evidence of a volcanic explosion. Fortunately, we have just such a gauge: for this investigation, our climate tool of choice is based on ice core analysis. Ice core analysis is a physical measure of climate that has the marked advantages of being very precise in dating specific events and also of being very reliable in determining past temperatures and precipitation amounts. Other natural measures of climate—like tree rings, which we'll discuss a bit later—don't go far enough back in time to have recorded the Toba event.

How does ice core analysis work? Over the course of a given year, snow accumulates when it falls onto a surface. During the summer, some of that snow may start to melt and then refreeze as winter starts. The new year's snow falls on that thin lens of last year's refrozen ice and accumulates. The new year's top snows, in turn, begin to melt in summer, and then refreeze. Over the course of several years, we begin to see a distinct layering as one new layer forms each year on top of the old. In addition, as new snow falls onto the old snow, a gradual transformation occurs in the old snow itself—the compaction by the new snow gradually forces out air from the older layers. The material no longer is snow but instead is in a strange, "slushy" transitive stage between snow and ice that is called *néve.* Over a long period—sometimes as long as twenty-five to fifty years—the néve continues to be pressed by the new snow above, which forces more air out, and it eventually becomes glacial ice.

Glacial ice is not normal ice, such as what you find in your refrigerator. First of all, because most of the air has been forced out, it is much denser—heavier—than its icebox equivalent. A three-inch-by-three-foot tube of glacial ice may easily weigh twenty-five pounds, whereas refrigerator ice of such dimensions might weigh only half that amount. Second, because of its density, light doesn't pass through it as easily as refrigerator ice. Glacial ice tends to have a darker, more bluish, color. But, most important, glacial ice still has those individual layers of the original annual snows embedded within it—and those layers (and their contents) are critical to the work of climatologists.

By the way, extracting that glacial ice can be among the most laborious and expensive science operations to conduct. I was involved in the drilling of a couple of ice cores in Antarctica for the Ohio State University Byrd Polar Center a few years ago. The process is similar to oil drilling—except that we drilled through the ice using a hollow-tube, three-foot-long, diamond-tipped drill so that the glacial ice was forced up and into the short tube. Then the operation was stopped, we brought the tube to the surface, the core was pushed out of the tube, the drill reassembled, sent back down the hole, and we drilled a new three-foot section. When the whole process was finished (filling boxes and boxes with the extremely precious ice), we packaged and shipped the individual sections of ice core to a laboratory in Ohio for further analysis.

When I was discussing the slow transformation from snow to glacial ice in an ice cap, I left out one important fact—if something falls onto that snow, it also becomes buried along with that snow by subsequent snows. That fact is extremely vital for the study of vol-

A section of an ice core taken near Siple Station, Antarctica. This section would date back to roughly 1500 CE.

canoes. We can literally see the ash deposits of an individual volcanic eruption embedded in the layers of an ice core. Consequently, we can precisely date the eruption and examine the chemistry of the ash to determine exactly what types of material the volcano emitted.

The only remaining possible obstacle to using ice core analysis for study of disasters like the Toba event is how far back the event occurred. Do ice cores go far enough back in time to see something like the Toba eruption of seventy-three thousand years ago? The short answer is yes. Ice core scientists in both Greenland and Antarctica (the earth's two present ice sheets) have actually drilled to the bedrock under those two landmasses—literally miles deep in the ice! That means, specifically for Antarctica, drilling over a mile into the ice (over 3,200 meters). The extracted ice core dated back 740,000 years and reveals as many as eight distinct glacial cycles![6]

So does ice core analysis show the Toba event? When we examine the ice core record from seventy-three thousand years ago, we see a distinct ice layer that contains one of the largest amounts of volcanic sulfur present in the entire ice core record.[7] That would indicate that the Toba eruption was big—very big—and that it contained that critical climate-cooling element, sulfur. The likelihood is therefore that massive amounts of sulfur were ejected high into the stratosphere when Toba exploded seventy-three thousand years ago. Also, the ice core record suggests that the Toba eruption lasted not for minutes or hours but for days—perhaps as many as six full days of continuous explosive eruption. Without a doubt, Toba had time to eject megatons of material into the air.

But that doesn't tell us how cold the weather was after the Toba eruption—and for how long. Intriguingly, ice core analysis can even tell us specifics about how cold the weather was—but to do that we need to learn a bit more about ice core analysis. Detailed ice core analysis occurs in the laboratory. For the cores that I helped to drill, that analysis was conducted at the Byrd Polar Center at Ohio State University in Columbus, Ohio, under the guidance of world-renowned climatologist Dr. Ellen Mosley-Thompson. There the ice cores were taken into a "clean" room, isolated from the outside—so that no risk of contamination can occur—and samples were taken from the layers for detailed isotropic analysis. To understand the usefulness of isotropic analysis in climate research, we need to discuss basic atomic theory.

Everything around us is made up of atoms. Water, for example, consists of atoms of the two elements hydrogen and oxygen. Each molecule of water is composed of two atoms of hydrogen bonded through an atom of oxygen. All atoms of an element have essentially the same chemistry, but they sometimes have a variety of different masses, or isotopes, which relates to the number of neutrons in the nucleus. For example, there are three stable isotopes of oxygen with masses 16, 17, and 18 (written ^{16}O, ^{17}O, and ^{18}O, respectively). In each of these cases, the lighter isotope (^{16}O) is by far the most abundant in nature, so that most water molecules consist of two hydrogen atoms and a ^{16}O atom, although there are small amounts of water with two hydrogen atoms and an ^{18}O atom. The third isotope of oxygen, ^{17}O, is radioactive, meaning that it rapidly decays to other elements. This isotope is rare in nature.

The key to using this concept of different isotopes of oxygen involves the determination of the relative amounts of ^{16}O and the less abundant ^{18}O. Scientists have found experimentally that as one travels to successively colder regions, the rain or snow becomes progressively depleted in the amount of heavier ^{18}O isotope that it contains (basically that heavier oxygen rains out first, leaving the lighter ^{16}O still in the cloud). This depletion is directly proportional to the temperature difference between the place where the water originally evaporated and the place where it eventually rained out. Because there is more evaporation in the equatorial regions than precipitation and the reverse in the higher latitudes, water vapor normally is being moved poleward. Thus, if one measures the ratio of ^{18}O to the amount of ^{16}O in rain or snow at any location, one can determine the temperature between that location and the equatorial tropics.

But the great usefulness of this concept to climatologists comes in realizing that, at any given location, if the climate didn't change, then the ratio of ^{18}O to ^{16}O in snow (and then ice) should be constant throughout time; thus, if we dig up an ice core, the core should have a constant ratio of ^{18}O to ^{16}O. But if climates have changed in the past, the ratio of how much ^{18}O to ^{16}O is in each layer can tell us very precisely how the temperatures of that location have changed over time. So laboratory analysts carefully measure the amount of isotope ^{18}O in each layer's ice as compared to the amount of isotope ^{16}O. When we plot that ratio on a graph against time, we suddenly have a road map of climate change going back over thousands of years.

What does the ^{18}O /^{16}O ratio tell us about conditions on the earth that followed the massive sulfur deposits seventy-three thousand years ago? According to laboratory analysis, the eruption of Toba was followed by decades of some of the lowest ice core oxygen isotope ratios in the last hundred thousand years. In other words, following Toba's explosion, the world experienced for a short time temperatures colder than even during the height of the last ice age some twenty thousand years ago. Temperatures across the globe perhaps plummeted by as much as 16°C (24°F)—particularly in regions away from the equator.[8] Human populations living in Europe and northern China would have been completely eliminated.

Even in more temperate zones, conditions would have been unbelievable. The region of the present-day Middle East and India would have suffered dramatically from the ashfall and from the volcanic-induced cold. It is clear that, given the traumatic climate change, the surviving remnants of humanity had to be located in a region where the massive climatic effects of Toba were minimized as much as possible. Such a region would most likely have been in very isolated tropical pockets in eastern equatorial Africa.

More traumas built on the catastrophe. As the bitter cold deepened and sunlight waned, plants died. For example, there is some evidence of massive deforestation in Southeast Asia after the eruption of Toba. Encouraged by a worsening climate, dwindling food supplies, and coupled with inadequate clothing and housing, two more evils, famine and disease, would likely have impacted entire segments of humanity at the time.

Eventually, with only small isolated groups left in equatorial Africa, survival may have depended on cooperation between remaining groups. Groups unwilling or unable to cooperate and adjust fast enough to the changing conditions would simply have died out. The surviving human populations would have continued to adapt and slowly begun to recover.

How low did humanity's numbers sink? There is no clear answer. Some scientists have suggested that it was as few as only forty to six hundred females—or a total human population of fewer than three thousand persons. The highest estimate so far has only been about ten thousand females of reproductive age making it through the bottleneck period of the Toba eruption. Yet, even if we accept the highest estimate, the entire human race at the end of the Toba event may have numbered no more than the residents of a single small town today.

Is there some lesson that we might take away from Toba's near-genocide of humanity? Unfortunately, yes. While none has erupted since Toba's explosion some seventy-three thousand years ago, the world still contains a number of similar supervolcanoes.[9] One in particular is quite worrisome. Unknown to most people, most of the beautiful Yellowstone National Park in Wyoming is the central caldera—the volcanic opening—of one of the most massive supervolcanoes yet identified by scientists. Yellowstone is a supervolcano bigger than Toba. All of Yellowstone's hot springs and geysers—even Old Faithful itself—exist because of an incredibly massive magma chamber located underneath the park.[10] If the Yellowstone supervolcano were to erupt, many scientists believe that it would have similar—or perhaps even larger—impacts on the earth than did Toba's eruption of seventy-three thousand years ago.[11] Some scientists believe that a supereruption of Yellowstone could conceivably end modern civilization. And we do know that Yellowstone has erupted in the past on a fairly regular eruption cycle of six hundred thousand years.

The last megaeruption of Yellowstone was 640,000 years ago . . . are we overdue for another supereruption? And, if so, would the human race this time be pushed beyond Toba's destruction, to the total extinction of humanity?

Time: Approximately seventy-three thousand years ago
Location: Region of present-day Ethiopia, Africa

Aslona watched with a mother's natural concern as her children played along the banks of the river. Long shadows were appearing from the riverside trees as the day globe began to disappear over the western horizon. Life had been hard since the monstrous big noise that her tribe's storyteller sang of many seasons ago. So many of her people had died from the sudden cold and through the illnesses—but now the gods were smiling once again on her people.

Food was once again becoming plentiful. The tasty red berries on the brushes, the juicy fruits on the trees . . . and the men had become once again successful in their hunting of the antelopes and other game. The tribes of her small river valley had even begun to work together. She glanced down at the animal skin moccasins on her feet. That had been a good idea that the clan on the other side of the river had had—these skins protected the feet against hard rocks as well as kept them warm in the cold nights.

The deep reddish light of the day orb colored the high, fast-moving clouds in an array of dark purples and violets. A cold breeze whistled down the river valley.

It would be dreadfully cold again tonight, Aslona realized. Already the men of the tribe were building a large warming fire in the open area among their huts. Many of the older members of the tribe were shuffling over to it so that they could sit by the crackling blaze, warming their cold bones.

"Toka! Malsa!" she called to her children.

The two—a girl of five seasons and a boy of three—

raced up the path from the river toward her. She embraced both lovingly as they leapt into her arms with happy squeals of joy.

A broad smile crossed Aslona's lips. As long as her children were safe and fed, she knew that she—and her tribe—would survive.

Chapter 4

THE MYSTERY OF THE SAHARAN HIPPOS

One of the perplexing aspects of climate change is that we humans have a tendency to dismiss climate as not having a major influence on our lives, yet we don't have the same difficulty when we consider its effect of animals. If we uncover fossils of a certain animal species in an area where it currently doesn't live, one of the first things that most people use to explain the absence is climate change.

Consider, for example, the intriguing mystery of the Saharan hippopotami.

Time: Approximately 5300 BCE
Location: The Wadi Mathendush, central Sahara Desert, in modern-day Libya, Africa

Yes, the hunt had been successful and the tribe would eat well tonight and tomorrow. Yet Tohar nodded sorrowfully to his hunters as he brushed some of the

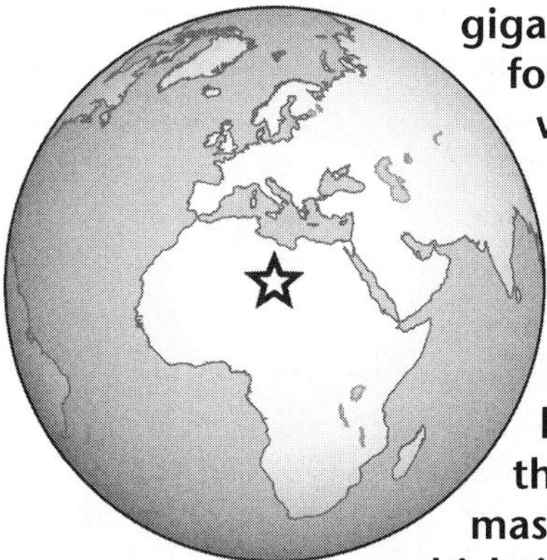

gigantic water beast's blood from his forehead. Celebration was mixed with sadness. On the whole, his men had done well. As head of the tribe, he had earlier been concerned about the hunt's success. After all, this had been one of the first times he, as the new leader of the tribe, had resolved that they should hunt one of the massive, thick-skinned water beasts, which inhabited the marshlands downstream from their camp. But despite the careful participation of all of the tribe's hunters, the hunt had also proven ultimately costly to the tribe and to Tohar in particular. Young Jocam, his brother's son, had been badly injured when the maddened beast had thrashed about in its final death throes and crushed the youth's right leg. Even with Garn's care, it was likely that Jocam would die during the night. Yes, the large water beast had given his men a hurtful fight but the massive amount of meat from the beast would fill the bellies of his tribe for some time.

So, despite the injury to Jocam, the hunt had been a godsent bounty. Tohar frowned, shifting from his concern over Jocam to the health of the tribe. Yes, the successful hunt had only been accomplished through the gods' providence. Perhaps it would be wise to honor those gods who had given the tribe this great gift. He gestured toward Garn, the tribe's shaman. Garn was different from the rest of the tribe since he spoke with a strange lisp and one eye did not quite follow the other. But was that not a sign of the gods? After all, did not the gods bless Garn with the ability at times to cure those of the tribe who suffered injuries and illnesses?

As Garn approached Tohar, the leader pointed toward the stone bluffs overlooking the river. "Go," Tohar commanded, "mark the rocks with the memory of this great water beast!"

In what is now the middle of the Sahara Desert there are hundreds of ancient rock drawings and paintings that have existed for thousands of years (and are still visible today). These rock murals carefully pecked or painted by ancient artists are filled with a menagerie of fascinating—and easily identifiable—animals. Rock walls in the central Sahara Desert display the ancient images of elephants, buffalo, antelopes, giraffes, hippopotami, crocodiles, and rhinoceroses. Interestingly, these animals are often drawn in elaborate scenes in which hunters in canoes are pursuing the seemingly exotic game. Canoes and hippos in the Sahara? How can that be?

The study of rock art—and, in particular, the dating of those images—is a fascinating subject and one that is critical to the study of climate change. The vital question that a climatologist poses when examining rock art—especially when it is of plants or creatures that don't currently exist in that area—is "How old is that image?" Was it carved with cutting tools, pecked with hammerlike implements, or painted with crude brushes just a few hundred years ago or a few thousand years ago?

The answer to that question can help determine what type of climate existed when that rock art was first put onto the stone walls. So how can we determine the date? Sometimes, archaeologists can link the images to traces of human existence in the area—pottery, bones, and so on. Sometimes, the style of the artwork might tell us about the date of the image—some types of images or some patterns found in the rock art are linked to specific peoples. But one of the best means to date scientifically the rock art is to use laboratory analyses.

I have become perhaps a bit more knowledgeable than most climatologists on that subject of scientific dating of rock art, since I am married to a rock art chronologist and conservationist (and, in my humble—totally unbiased—opinion, one of the best).[1] Rock art chronologists have the daunting task of establishing scientifically the dates for when these mysterious images were first painted, carved, or pecked into the rock.

There are a number of ways to date rock art.[2] As I mentioned, one can link it to nearby dated archeological ruins (and assume the two were made at the same time) or one can associate it with ethnographic knowledge of given cultures (for example, a people whose times of existence we have established might have used certain consistent symbols). But, often, direct dating of the rock art itself is the best means of establishing its relative age. The tool of choice for one of the world's foremost rock art chronologists and a good friend of mine, Dr. Ron Dorn, normally involves the radioactive decay of certain trace elements.[3] And now we must sidestep from our mystery for just a moment to learn about this specific type of investigation so that we can apply it to our climate mystery.

As I mentioned in the Mystery of Humanity's "Near Extinction," everything around us consists of atoms. All atoms of a particular element have essentially the same chemistry, but they sometimes have a variety of different masses (isotopes) relating to the number of neutrons in the nucleus. For this mystery, we need to discuss the element that makes up living tissue, and that is carbon. Carbon normally exists with an atomic mass of twelve—twelve protons and neutrons in its nucleus. However, two other forms of carbon can also exist: ^{13}C with one additional neutron in its nucleus, and ^{14}C with two extra neutrons. In nature, ^{12}C is by far the most abundant form. ^{13}C is an isotope that is termed "stable"—that is, it doesn't decay into other elements. For this mystery, the key isotope is ^{13}C, the radioactive, or unstable, isotope of carbon.

With unstable isotopes such as ^{14}C, the atoms undergo spontaneous radioactive decay by the loss of their nuclear guts (such as protons or neutrons, or even, in some cases, their outer electrons) and, as a result, they may transmute into a new element. ^{14}C, for example, decays into stable nitrogen (^{14}N). The critical factor for our mystery is that the decay rate—the time it takes to change ^{14}C into ^{14}N—makes for a very consistent environmental clock. In other words, we know exactly how long it takes for a quantity of radioactive carbon to change to a specific amount of nitrogen. This natural clock is the basis of all radioisotope-dating methods.

Scientists take a precise measurement of the present isotope concentration and use that to indicate the amount of time that has elapsed since the sample began to decay. The term that scientists use to refer to the amount of time that it takes for an isotope to decay to half of its original amount is "half-life." In the case of the ^{14}C isotope, the half-life is approximately 5,730 years. Thus a plant that died 5,730 years ago currently has only half of its original ^{14}C content remaining in it. After 5,730 more years from today, it will have only half as much again, that is, 25 percent of its original ^{14}C content, and so on.

A given isotope must have several specific characteristics if we want to use it for climate research. First, the isotope, or its decay materials (which scientists term "daughter products"), must occur in measurable quantities (that is, we must have enough to sample), and its rate of decay must be measurable (for this climate change research, we want a decay rate measured in years as opposed to, say, nanoseconds). Second, the isotope must have a half-life that is appropriate to the dated period (a radioisotope with a half-life of tens of years isn't good for looking at events hundreds of millions of years ago). Third—and this is somewhat tricky—the initial amount of the isotope must be known or estimated. And fourth, there must be some connection between the event being dated and the start of the radioactive decay—normally, for radiocarbon dating, that date is when the plant or animal died. Why is that the case? When we are living, we are constantly replenishing the amount of radioactive carbon in our body through processes such as simple breathing. When we die, the amount of radioactive carbon becomes "fixed," that is, it doesn't change anymore—and the radioactive decay clock starts ticking.

With that as a bit of background about rock art dating, let's return to our mystery of the Saharan hippos. As I mentioned, hundreds of ancient rock drawings and paintings of hippos, elephants, even giraffes, exist in what is now the middle of the Sahara Desert. Through the use of isotopic dating, the creation of these rock art images conservatively dates back six to eight thousand years ago (4000–6000 BCE).[4]

Okay, these rock art images are old—indeed, very old—and they depict an environment nothing like what currently exists in the middle of the Sahara Desert. So here is the mystery: How could hippos, rhinos, and crocodiles exist in the middle of what today is the world's greatest desert?

The answer is simple: climate change. And in this case, a climate change that was definitely not created by people. Actually, it's the reverse: the peoples in this mystery were directly impacted by the climate change—they didn't cause it. That's a very important thing to note—a point that sometimes gets lost in today's headlines involving global warming. Climate change—most of the time not caused by humans—has always happened! Change in climate happens all the time. Sometimes it is the result of our activities, but most of the time, it is not. Climate has changed in the past, it is changing now, and—whether or not we cause it—it *will* change in the future.

But to determine what climate change happened to create such a different environment in North Africa eight thousand years ago, we need some additional investigative tools. In this case—in a similar fashion to the Mystery of Humanity's "Near Extinction"—we can look at the oxygen isotopes in ice cores from Greenland and Antarctica. But, for this mystery, we also have other tools as well. One very important means of establishing past climates at the end of the last ice age is through the study of deep-sea marine cores.

Occupying more than 70 percent of Earth's surface, the oceans are a very significant source of paleoclimatic information. Essentially, this is because the ocean bottom is one of the world's greatest garbage pits, the eventual home for billions of tons of dirt—and the dead bodies of trillions of microscopic sea creatures. Indeed, a huge amount of dirt and sediment (between six and eleven billion metric tons!) falls into the ocean each year. That sediment eventually sinks to the bottom, slowly accumulating on the ocean basins in thick layers. Just like the snows whose slow transformation into layers of an ice core can tell us of past climates, so too, these layers of accumulated sediment at the bottom of the oceans may be indicative of climatic conditions.

For the study of the sediments at the bottom of ocean basins, termed "deep-sea marine sediment analysis," the key is the exact composition of the layers. The ocean sediments are, in part, composed of the remains of planktonic (near surface-dwelling) and benthic (deep-water) organisms that provide a record of past climate and oceanic circulation. The remains of thousands of these organisms, called *biogenic oozes*, are primarily composed of microscopic skeleton shells.

Intriguingly, the very character—the concentrations, the size, the shape, and even the appearance—of these minuscule shells can tell us a tremendous amount of information about the climates that existed when these tiny sea creatures lived and died. Some of the most important of these minute marine animals are *foraminifera* (a form of zooplankton) and the much smaller shells of *cocolithophores* (unicellular algae). Let's examine some of the types of climate information we can glean from deep-sea marine cores.

In one of the first climate analyses using these microscopic animals, a scientist by the name of Schott recognized as far back as

1939 that the variations in the number of certain foraminifera was indicative of alternating cold and warm intervals in the past.[5] That means that, if we identify differences in the numbers of these foraminifera through the layers of ocean sediment, we can start to determine what the temperatures were when they were alive.

Second, numerous studies have shown that shape differences in certain tiny sea creatures may result from changing environmental conditions. Perhaps the best known of these is the difference in coiling directions of the tiny shells of particular species of foraminifera. Researchers have identified a clear relationship between water temperature and actual "coiling direction" of the minute shells of these creatures. Think, for example, of the coiling that you see in a giant conch seashell. Now picture that shell as a microscopic fragment that is substantially smaller in size than the head of a pin. What you might discover is that "dextral coiling" (that is, right-hand coiling) of those tiny sea creature shells predominately forms in warm tropical and subtropical waters, while "sinistral coiling" (or left-hand coiling) occurs in colder waters. So, if we study the layers of a particular deep-sea core and identify the changes in the number of individual microscopic shells exhibiting coiling in a particular direction, we may have a record of how the surface water temperatures have changed over the course of time. In fact, for some species, we can be very precise about the temperatures: for one species of foraminifera, the predominance of one coiling direction or the other occurs when the ocean temperature is either above or below 10°C (50°F) in the North Pacific. In essence, these tiny sea creatures become microscopic biological thermometers.

Another manner in which foraminifera can act as biological thermometers is in their actual size. For some species of these tiny sea creatures, warmer waters are linked to a larger shell size while individual foraminifera fragments in colder waters are much more compressed.

Finally, since all of the skeletons or shells of those microscopic marine animals are composed of calcium carbonate—calcium linked to oxygen—the amount of oxygen-18 can be determined. Recall that ^{18}O is the isotope of oxygen whose concentrations in rain and ocean water are temperature dependent.

And what do the painstaking laboratory analyses of all of those deep-sea marine cores tell us?

Fundamentally, they reveal to us that cold has been the planet's norm over most of the last two million years. The climate for most of those two million years has been much, *much* colder than the present day. Huge ice sheets covered much of northeastern North America and northern Europe. Massive mountain glaciers flowed down into the plains of most continents, but particularly in Europe and North America. Sea level was some four hundred feet lower than in present-day—meaning that Florida, for example, was twice as big as it is today and that the area between North America and Asia, today an ocean strait, was, for most of the last two million years, dry land.

Those frigid periods, marked by the existence of huge ice sheets, were part of long-term pulses of cold that were likely the result of changes in Earth's orbital character. One very interesting aspect of those advances is their duration. These long-term cold periods, when temperatures were globally five to eight degrees colder than those of the present day, lasted for exceedingly long periods of time—up to a hundred thousand years. Over the last two million years, we have had perhaps as many as twelve separate ice age periods. And between those cold periods were much shorter warm periods, called interglacials. Each of these warm interglacials lasted about ten thousand years. Indeed, we right now are living in an interglacial that started ten thousand years ago.

Ten thousand years is a critical number. It means that *all* of recorded civilization—from before the Egyptian Empire to today—has occurred during the relatively short time period between long ice age episodes. Additionally, the average length of an interglacial—ten thousand years—has critical implications for our future climate as I'll discuss later. For now, with regard to the Saharan hippos, we need to realize that ice cores and deep-sea marine cores tell us that, at the time the hippos were being sketched on the rocks in North Africa, the earth was coming out of its last great ice age.

During that last ice age, many of the arid places around the world were much wetter than they are in the present.[6] For example, what today is the Great Salt Lake of Utah was, back in the last ice age, a very small part of a much larger lake called Lake Bonneville. That ice age lake covered nearly twenty thousand square miles, making it nearly as large as Lake Michigan. As the world came out of that ice age and warmed, Lake Bonneville slowly has evaporated away, leaving just a few small remnant lakes, of which the Great Salt Lake is the largest remaining.

Similarly, during the last ice age, Lake Chad in North Africa was much bigger than it is today. Indeed, what is today's Lake Chad is simply a very small relic of a huge inland sea, what some have termed "Megalake Chad." At its largest extent, this inland sea is estimated to have covered an area of 150,000 square miles—a larger area than that of the world's largest present-day freshwater body, the Caspian Sea. Over time as we have progressed further into this current interglacial, that inland sea has gradually evaporated away. Today, Lake Chad is only about five hundred square miles—due in part to high irrigation demands on the lake as well as recent reduced rainfalls. Indeed, twice in recent times—in the early 1900s and again in the 1980s—Lake Chad has completely dried up.

The water for Megalake Chad of the past was the result of substantially increased rainfall. In North Africa during the times of Megalake Chad, it rained hard and often. Such rainfall occurred because of significant changes in the world's *general circulation*, or the fundamental wind patterns of the atmosphere. The world tends to have very distinct zones of weather based on two important attributes of the world's winds—first, the need to balance hot and cold by having hot flow toward cold and, second, the changes created in wind direction because of the spinning of the world (something called the *Coriolis effect*).

These zones generally are relatively stable and well identified. They include: a cloudy, rainy region near the equator called the *intertropical convergence zone*; a large area roughly between about 5° and 25° latitude called the region of the "trade winds" ("trade" because it was the region that colonial ships sailed through in order to have the best, most persistent winds for crossing the Atlantic Ocean).

Poleward of the zone of the trade winds is a relatively hot, dry, and windless region between 25° and 30° latitude called the "horse latitudes." This region—dominated by high pressure (leading to clear skies and warm, dry air)—was labeled the horse latitudes by early explorers of the New World. According to legend, colonial-era sailors, who were becalmed in this windless zone, reportedly threw their cargo of horses overboard to lighten the ship in order to continue their travel. Because the horse latitudes are associated with the warm and dry air of high pressure, it isn't surprising to find that most of the great deserts of the world occur in this zone.

Continuing poleward of the horse latitudes, we enter a region of consistent winds blowing from the west (found between 30° and 60° latitude) called the "westerlies." This is the region that includes most of the United States and Europe. It is characterized by numerous winter-time storms moving from west to east. Concluding our wind tour of the planet, we find a relatively consistent cloudy and stormy region around 60° to 65° latitude and, finally, a region of cold winds (generally blowing from the east) poleward of 65° latitude.

As I mentioned above, generally these zones are relatively stable—they don't move much from year to year. However, they have changed fairly dramatically between the time of the last ice age and modern times. In particular, the region of the horse latitudes—that zone of hot, dry, and windless weather, in which today we find most of the world's deserts—was likely located significantly farther south in the Northern Hemisphere during the last ice age. But what would that do to the world's weather?

The world's winds are all interconnected—if you change the location of one zone, all of the others have to adjust. Consequently, if the zone of the horse latitudes (and the location of the world's deserts) shifted southward during the last ice age, then the zone of the westerlies (and all of the winter storms that we experience across North America and Europe) also shifted southward. And that means North Africa—with the horse latitudes more consistently south of the region back then—was able to have much more rain than it does today. More water led to a more diverse ecosystem—which included hippos, rhinos, and even giraffes.

A brilliant geographer by the name of Karl Butzer was one of the first to study the geographic distribution and survival of some of these odd animals depicted on the rocks in North Africa.[7] He found that large herds of giraffes and elephants survived in the region until about 2900 BCE—indeed, the elephants that were used to attack Rome by the famous Carthaginian general Hannibal (from what is today Libya) likely composed one of the last isolated herds remaining in the increasingly arid North Africa. Over the time from 5000 BCE to 2900 BCE, the large-scale circulation patterns slowly shifted northward from their southerly glacial positions until they reached their present-day positions—which led to the creation of the large deserts of North Africa, as well as deserts in the North American Southwest.

There are some who have suggested that the birth of civilization around large, enduring rivers such as the Nile and the Indus—the Egyptian Empire as well as the civilization of the Indus Valley—was, in part, due to the contraction of habitable lands in that great swath of developing desert. Those lands were continually under the horse latitudinal high pressure as the world warmed out of the last ice age. But we'll save those studies for a later mystery.

Time: Approximately 5300 BCE
Location: The Wadi Mathendush, central Sahara Desert, in modern-day Libya, Africa

Garn lay crouched on the stone bluffs overlooking the river, studying his artwork. He had never before tried to capture the spirit of the great water beast on the eternal rock. Yet his lines had been true—undoubtedly the spirit of the water beast itself had helped him in his task. He had depicted with eerie accuracy the image of the massive animal on the rock surface—the huge round body with short, stubby legs and the mammoth head with its massive mouth and nose. One would have imagined that such a large and lumbering beast would have been easy prey—but in the water the thing was a demon, thrashing with wild abandon as the arrows pierced its thick hide.

In the end the hunters had been successful—Tohar's tribe would have food. Garn nodded in satisfaction at his portrayal of the water beast. The gods would be pleased—memory of this great hunt would last forever. A booming crack of thunder resounded over the area. Garn glanced up at the billowing clouds of a thunderstorm building over the area, and then with a final appreciative glance at his artwork, headed back toward his tribe's encampment.

Chapter 5

THE MYSTERY OF THE VANISHING HARAPPANS

A good climatologist, anthropologist, or sociologist should always ask two basic questions before asserting that climate change caused a civilization's collapse. The first is . . . well, let's hold off on that first question for a moment because the second query is the one that is posed most often. The fundamental *second* question about climate change and societal collapse is simply: *How* could climate change contribute to the demise of a great civilization?

That intriguing question has been asked a number of times in the past by archaeologists, sociologists, and climatologists, and it is one that is being asked often today with regard to the discussion of long-term effects of global warming. It is an important question that has its roots in an old concept called *environmental determinism.* The early proponents of environmental determinism, which is sometimes referred to as climatic determinism or geographical determinism, advocated that the physical environment exerts an absolute control on society and culture.[1] In essence, environmental determinism says,

"You—and your society—exist only because of your specific climate and environment."

Environmental determinism played a big role in science in the first half of the twentieth century. The central tenet of environmental determinism was that the physical environment of a society, particularly its climate, exerts a profound and deep-set control on the overall behavior of the society. Early anthropologists and geographers, for example, proposed (repulsively to modern thinkers) that the inhabitants of warm and relatively mild tropical climates had tendencies toward laziness and promiscuity, while the frequent variability in the weather of the middle latitudes created more industrious and goal-oriented inhabitants. These early scientists theorized that, because human biology changed very slowly in the face of these changes, it was crucial to trace the migrations of various groups from one land to another as to identify the underlying environmental conditions of the area from which they originated. A person with ancestry from Europe or Africa, according to these scientists, retained the deep-set climate-induced tendencies of those original locations.

As you may understand, this easily led people in the first half of the twentieth century to cite environmental determinism as a "scientific" means of justifying prejudice and racism. As such, the theory rapidly fell into general disrepute by the 1950s. Today very few—if any—scientists advocate such a hard-line "climate controls culture" posture.

However, intriguingly, an almost diametrically opposed mind-set is in evidence today. That mind-set states that human actions—either by design or accident—alone fundamentally control our environment and particularly our climate. In the minds of many today, natural variability is no longer considered to be the means by which our weather varies from year to year. For example, inevitably, the first question that I am asked when I am interviewed about any new weather-related disaster is: "But isn't this an indication of global warming?"

I must admit that I have severe problems with both views—environmental determinism, and what might be called *anthropogenic determinism*. First, with regard to environmental determinism, I believe that, because we all, as humans, do have intelligence, the concept of our climate and environment actually controlling our biology and society is generally repugnant to me. I like to believe that we're smarter than that. For example, as I will demonstrate throughout this book, past human cultures have generally found novel and effective ways to adapt to changes in their environment. Irrigation, new crops, and new methods of farming are but a few of the ways that various societies adjusted to environmental change.

However, even given our vaulted intelligence, environmental change can create or add to an underlying dissatisfaction in a society—dissatisfaction that might, if unchecked, eventually lead to social problems, revolt, migration, and perhaps collapse. I would imagine that, if a crop consistently fails and one's stomach is empty, such a person is perhaps a bit more receptive to thoughts of changing governance or of even moving out of the region. So I think we should be perhaps very careful before *completely* discounting the influence, but not "control," of climate and environment on society.

In a similar fashion, I think, we should be wary of assuming that the human race is now the sole cause of global climate change, a concept that I've called anthropogenic determinism. Undoubtedly, humans do have an impact on the climate. For example, research has conclusively shown that a city's roads and buildings do exert a strong and easily identifiable impact on weather.[2] A city's core is warmer than its periphery—sometimes as much as 12°F warmer! And on a larger scale, as I showed in the Mystery of the Weekend Rains, a colleague and I established that the region just offshore of the East Coast is rainier on weekends than on weekdays.[3] We theorized that, if rain in that area follows a seven-day cycle, the cause for that rain must be human-made. Why? The seven-day workweek is an artifact of humans, not nature. Anything showing a seven-day cycle in nature is very likely to be strongly influenced by people. In that case, we suggested that the human cause of the rainier weekends is likely the extensive pollution emitted from the East Coast metropolitan areas.

But, while we do exert an influence on weather and climate, we should realize that our impact is *not* the only influence. Without question, other aspects of nature still exert strong influences on our weather and climate. Variations in solar output, changes in ocean circulation and temperatures, and oscillations in the winds of the upper atmosphere among other factors have all been shown to affect and change our weather and, consequently, our climate. So, just as researchers backed off the absolute "climate determines society" paradigm in the last century, we should be equally careful of embracing the absolute "society determines climate" theory.

With that introduction to the interdependence of climate and social changes, let's return to the two fundamental questions that should be asked before concluding "climate destroyed a society." As I mentioned at the start of this mystery, the very first question that one should ask about how climate change contributes to the death of a civilization is, interestingly, very seldom asked by any researchers of climate change. If, as the second question implies, climate change can influence or even cause a civilization's fall, then what about its corollary?

In other words, as a first question before talking of collapse, let's ask, "Does long-term climate *stability* contribute to the initial growth and development of a society?"

We tend to forget that climate and weather are 24-7. Variations—and, just as important, consistencies—of weather and climate influence our civilizations and us all of the time. Just as some say that a given climate change, such as prolonged drought, may stress a civilization to its eventual destruction, doesn't it make sense that a given *consistent* climate may also initially set that civilization on the path to greatness? So when we study the popular question of how a climate change may have influenced a civilization's collapse, we should perhaps also address the basic question of what types of climate and weather initially allowed that civilization to develop at its beginning.

Yet if the question "Can climate help create civilization?" is important, why has it tended to be ignored? Partly, I think, it is because we as human beings are fascinated with *differences* rather than the *status quo*. It is much more exciting to hear how a single event like a drought, a storm, or a flood might have caused widespread death and perhaps the complete destruction of a given society rather than hearing about how a possibly long-lasting and peaceful set of weather conditions might have fostered the growth and development of that civilization.

Let's turn our attention to a climate mystery to see how that works.

One of the world's first great urban societies developed around the middle of the third millennium BCE. That prosperous society has been called the Harappan civilization or, sometimes, the Indus Valley civilization. It developed in the present-day contentious area of Pakistan and India and extended into the rich floodplains surrounding the Indus River and its tributaries (namely, the Jhelum, Chenab, Ravi, and Sutlej rivers).[4] The earliest archeological signs of this society date back to 4300 BCE in the form of small pastoral, agricultural-based villages.

However, over the next millennium, those villages began to consolidate and new villages formed throughout the whole region drained by the Indus River.[5] Soon a distinct urban society formed, such that its inhabitants had both the time and the human resources to develop new inventions such as the potter's wheel. Consequently, these growing villages continued to flourish over the next few centuries, and, as the Harappan people aggregated in urban areas, the growth of even bigger urban centers began to form.

Time: Approximately 4200 BCE
Location: Near the Indus River, in modern-day northwest India, Asia

Bélin straightened his aching back with a well-deserved sigh, rested on his hoe, and surveyed with a smile his newly sown field of grain. With the help of his two young boys, he had managed to plant a field that extended nearly forty paces by twenty paces of land—one of the largest stretches of tilled land in the region. He nodded in satisfaction. In a few months, that field would be filled with more wheat than he and his family could eat. And that meant Bélin could trade some of his grain for stronger oxen to pull his plow or perhaps even some jewelry for his wife.

Overall, life was good. However, Bélin knew from bitter experience that fortunes can change. The long and difficult trek so many years ago across the desert expanse from the mountains to the west had pushed him near death. Then after he and his then-pregnant wife had moved into this flat region, they had to survive through that rough, dry spell. Indeed, in the end, they had needed to relocate a few times before settling here near the Great River.

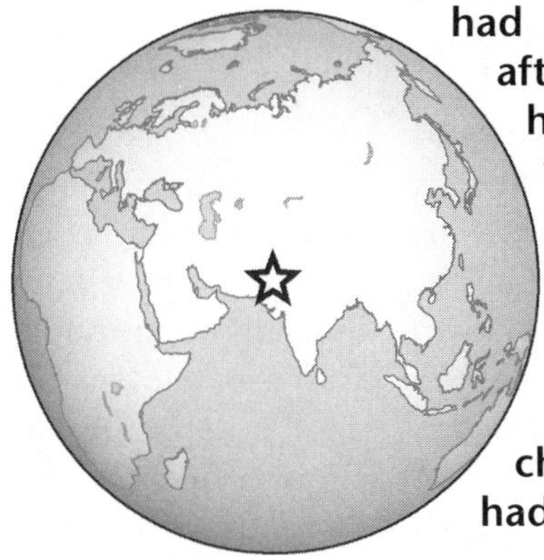

And now over the years, that choice had proven wise—the river had been good to them. It provided

unceasing water to drink and, equally important, continuous water for their crops.

Word had spread. Now, over recent years, more and more people had arrived in the region and a small community had blossomed, centered on the numerous tilled fields along the banks of the river. A few of the men had even turned their talents to making such useful items as pottery, taking wheat and barley in trade for their craft.

Bélin glanced at his youngest son. Old Urtho had noted that the boy had shown a good aptitude for pottery making and had suggested a couple of days ago that he could train the boy in the craft since Urtho had no children of his own. Bélin shook his head. It was a tough decision. The extra hands for working the fields were useful, but he could see that the finely crafted items Urtho made were becoming more and more popular. Yes, he would give Urtho's suggestion of apprenticing his boy some serious thought.

By 2000 BCE, or more than two thousand years after the fictional Bélin's time, this successful and thriving Harappan society had grown both in physical size and in the basic structures of the society itself. Physically, elaborate architecture was burgeoning with increasingly larger and more numerous buildings. Indeed, by the start of the second millennium BCE, some of the most populated cities of the time, such as the namesake metropolis of Harappa and its equally populous sister city, Mohenjo-daro, dotted the landscape of the Indus Valley. Culturally, Harappan civilization was becoming institutionalized with the creation of basic societal rules, for example.

Then, without any catastrophic volcanic eruption or asteroid impact, the Harappan civilization apparently disappeared within a few short centuries.[6] The once-lush, fertile area was reduced to a set of nearly abandoned or completely abandoned villages and a few sand-covered deserted cities. That desolate situation continued unabated across the region through the next couple of millennia, until the whole region was renamed—not "the breadbasket of human society," but instead the more stark appellation of the Thar Desert. This utter transformation of a region containing a prosperous and successful culture into a sparsely populated, hostile desert has led many scientists and environmentalists to cite the Harappans as some of the world's first poster children for the cataclysmic effects of climate change.

But is that a legitimate claim? Let's study both of our two fundamental climate/society questions and see what answers we may find. First, how did the society initially develop, and what role, if any, did climate and environment play in that development?

The great Harappan civilization originally developed through the interaction of nomadic peoples from the highlands of Baluchistan and southern Afghanistan with the sedentary agrarian populations in the west Indus Valley. Archaeologists tell us that the earliest farmers established, at first, small villages primarily near rivers, and, as populations grew, eventually founded some of the world's first large cities. The key question that a climatologist would pose is whether the creation of these urban centers was the result of any underlying climate of the region.

There are two basic lines of thought here. One theory states that the overall climate influence on the development of the Harappan civilization was positive, not negative. The proponents of this theory argue that the climate of the Indus Valley between 3500 and 2500 BCE became increasingly favorable (wetter) for settlement so that the peoples migrated into the region. As more and more people moved into the area and existing settlers had more children surviving, villages developed and then cities. This is the theory that nonarchaeologists generally have accepted. In particular, two famous mid-twentieth-century climatologists, Reid Bryson and Thomas Murray, supported this theory, first proposed by Gurdip Singh,[7] in their interesting book *Climates of Hunger*.[8]

The second theory for the initial development of the Harappan civilization is the opposite of the first. Its proponents—primarily archaeologists—say that urban centers in the Indus Valley grew because of increasing aridity and drought at the time of settlement.[9] This theory holds that drought would have led to fewer game animals and an increasing need for irrigation of crops from existing waterways. Consequently, people grouped nearer to the rivers and subsequently established villages and then eventually cities.

And, to be fair, there even is a third group who suggests a completely nonenvironmental viewpoint of Harappan cultural development.[10] These scientists argue that climate change was insignificant to the growth and eventual decline of Harappan civilization.

So which of these theories is correct?

I wish I could definitively tell you. But science doesn't always explicitly tell us things for certain. It gives us clues—sometimes too many clues—and then allows us to play Sherlock Holmes. What set of principles best fits the available facts? And, importantly, what facts do we have? Unfortunately, because the Harappans hadn't invented rain gauges (those came about three thousand years later!) or kept tidy annual logbooks of the weather, the best that we can do is examine the likely weather they experienced through inexact proxy evidence, such as pollen-based climatic reconstructions, using the science of palynology.

So let's be sidetracked for a moment from our Harappan mystery and take a closer look at what is one of our chief means for solving this great Harappan climate mystery, the science of *palynology*, or pollen analysis. My excellent contact for the details of this fascinating science is a very talented scientist who traverses the globe accompanied by her archaeologist husband in search of knowledge of past environments, Dr. Patricia Fall of Arizona State University.

Palynology is founded on several key principles. First, pollen grains and spores—the genetic building blocks of flowering plants and ferns—are extremely hardy and long-lasting. Their cell walls are made of a substance called *sporopollenin*, or what might be termed nature's plastic. In simple words, pollen grains don't easily decay under even the harshest climatic conditions. Second, a variety of surface features on pollen grains provide distinct patterns and, along with their shape, they can be linked to specific plant types and therefore are generally representative of a locale's basic vegetation. Third, although pollen production varies, pollen grains from many plants are often produced in vast quantities and are subsequently deposited by wind, rain, and running water. Finally, because they are very small in size, most being measured in microns, mere millionths of a meter, millions of pollen grains or spores can be found in a teaspoon-sized sample.

A central issue in pollen analysis is dispersion. Wind-pollinated plants produce large amounts of pollen that are distributed widely by the air, whereas other plants rely on animals to carry their pollen. These other plants tend to produce less pollen and to disperse it more locally than their wind-assisted counterparts. In general, most pollen deposition in forests, for example, reflects the surrounding local or regional vegetation with few pollen grains dispersed not more than a few hundred yards from their origin point. But sometimes, depending on weather conditions, pollen may be dispersed thousands of miles from its source, particularly in treeless vegetation. These factors must be considered when using pollen as a measure of vegetation changes for a given area. Depending on the plant species and the type of vegetation, pollen scientists generally assume that the pollen is reflective of the local area. But the essential element of pollen analysis is to realize that an individual type of pollen (and consequently an individual type of plant) does not exist in isolation; it is part of an overall group of plants that all are producing pollen over an area. An allergy specialist, for instance, can tell you that during a given pollen season, it isn't just ragweed that is in the air but perhaps hundreds of other types of pollen as well. Palynologists call this atmospheric soup of many different types of pollen an *assemblage*.

When the wind blows or rain falls, this pollen soup drops onto the ground. Of course, some of it falling on the land might be blown around again but some of it might also fall, or more likely be washed, into a lake. Pollen deposited into a lake can be one of the key indicators used to measure climate change. Eventually, deposited pollen assemblages will settle fairly uniformly on the lake bottom. Subsequently, the next year's pollen will settle on top of that, and so on, year after year. In some circumstances such as where the lack of oxygen prohibits mixing, these layers of pollen-concentrated sediment form annual layers called *varves*.

In terms of climate study, palynologists push or drill hollow core tubes into these lake-bed sediments. They then carefully extract those cores to obtain long tubes of the lake-bottom sediment material, and take them back to the laboratory. In the lab, technicians meticulously section the sediment cores into small intervals, each only a few centimeters thick. These sections are then treated with a series of acid baths to remove the inorganic material (rock and dirt) from the sediment, leaving only the pollen (this also reinforces just how tough those pollen grains are). The remaining pollen and spores are subsequently chemically stained and mounted on slides for microscopic analysis. Finally, the exact number of each type of pollen must be painstakingly tallied and recorded. Many researchers tend to use the

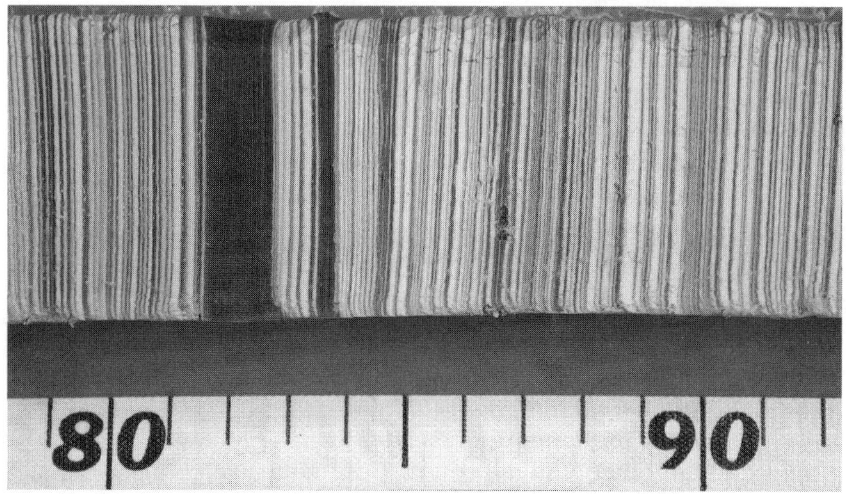

The laminations, or layering, of sediments (with pollen embedded and metric ruler underneath for scale) from a core taken in central Jordan. Photo courtesy of Dr. Patricia Fall of Arizona State University.

young eyes of their graduate students for this laborious and strenuous task. And, perhaps not surprisingly, I have noticed that a number of palynologists wear either contacts or glasses.

The end result of all of this work is the creation of a pollen diagram that relates the relative number of specific types of pollen as a function of the depth (and, hence, age) of the sediment core. The purpose of the diagram is to depict the gradual regional vegetation changes through the depth of the core (and, therefore, the age of the core) so that we can determine the changes in the environment and consequently the variations in climate that may have caused those changes.[11]

One of the fundamental concerns in pollen reconstruction analysis involves the rate of sediment deposition. We want to show changes in pollen type through time, not depth. So how do we convert the depth of a given sediment core to the corresponding time when the sediment and pollen were actually being deposited? In some cases, we can determine the underlying rate of deposition through knowledge of current rates of pollen deposition over the area and then assume that this rate has been relatively constant over time. But, probably, the best means of linking core depth and time is through radiocarbon dating of the pollen core at different layers.

Once the ages of the various layers of the sediment core are established, we can then study the changes in plant pollen that have occurred for that area over time—which brings us back to the Mystery of the Vanishing Harappans.

If we can record the changes in the types of plant pollen in the Indus Valley region from the present back to the time of the Harappans, we may be able to infer possible changes in climate. Fortunately, such records exist. The most complete record of pollen for the region is from several lakes and dry lakes (or *playas*), located in the central Thar Desert. Based on that record, a number of scientists suggest that the Indus Valley region at the end of the last ice age some ten thousand years ago was dry and arid. However, as the global climate warmed, this region began to get significantly wetter.[12]

In the Indus Valley sediment cores, palynologists have identified grass pollen flourishing throughout the region as our current ten-thousand-year warm period following the last ice age began, particularly with gradual increases in the number of marsh plants. Through the next five thousand years, there are indications of periodic droughts but nothing cataclysmic. By 5000 BCE, another wet period had begun, perhaps the result of more winter rains (as inferred from the reconstructed high lake levels in the region). This increased rain period was followed by the onset of a drier period sometime after 3000 BCE.[13] One slight complication to this pollen picture, however, comes from humanity.

People have a tendency to like certain plants and to take them along when they move. For example, I live in the Desert Southwest of the United States. At one time, Phoenix, Arizona, and similar Southwest cities were areas where hay fever and allergy sufferers migrated partly because of the lack of allergy-causing pollen in the air. I say "at one time" because now, unfortunately, the Desert Southwest has morphed into one of the country's worst regions for allergies. Apparently, when those thousands of allergy sufferers came here, they also brought their Bermuda grass for their front lawns, their olive trees because they flourished in the warm sun, and their orange and other fruit trees because of the excellent growing conditions. Consequently, the Desert Southwest with its light winds and low rain (both which lead to higher amounts of pollen in the air) is now full of alien forms of pollen. So it can often be, as I can personally attest, a miserable place for allergy sufferers.

I wonder if some future palynologist might wonder how all of that alien plant pollen keeps showing up in the pollen record of this desert region (when previously those plants hadn't been here). Remember, pollen scientists in general assume that the pollen in their samples is from local sources. Consequently, future climate researchers of the Desert Southwest might ask, "Was it climate change that caused all these different plants to flourish in this region?"

In a similar fashion, it is likely that some of the plants reflected in the pollen record of the Harappan region actually may have been brought there by peoples who moved into the area bringing their crops with them. The pollen of some economically beneficial but nonnative trees, for example, has been found at some Harappan sites. So a question can be posed that is a bit hard to answer: Were these new beneficial plants introduced by the Harappans or were they the natural result of transitions in climate that better allowed such trees to grow in these areas? Palynologists have to sort through which plants are native to a place and which are not.

With the potential caveat of native versus human-caused plant introduction, the pollen record of the Indus Valley appears to indicate that the region experienced a wet climate prior to, and then during, the initial settlement of the Harappans around 3500 BCE.[14] This wet period may have helped to promote the widespread adoption of crop cultivation and sedentary communities throughout the region. However, as the society took a foothold in the fertile plains of the Indus Valley, a drying trend (with possibly more frequent droughts) established itself over the region. That likely led to a shift of the increasing population to more permanent water sources, such as, in this case of the Harappans, the running rivers of the Indus and the Ghaggar-Hakra watersheds. These river settlements, in their more constricted geographic space, helped to promote a more urban lifestyle. Thus the Harappan urban civilization developed and flourished. At its height around 2500 BCE, this early society had more land under cultivation than in the Nile Valley and Mesopotamian civilizations combined.

And then the Harappan civilization of great cities began to decline—and eventually disappeared. What happened?

Part of the answer might be found in research by the science of *geomorphology*—an academic field that involves the investigation of the shape and flow of river systems. Studies from this field's scientists may give us particularly valuable indicators of how much water was available to these early peoples.

Geomorphologists tell us that a large number of Harappan settlements were located along a now-extinct river system called the Ghaggar-Hakra, which included the river and its tributaries that generally flowed parallel to, but southeast of, the Indus River.[15] The high number of settlements along this extinct riverbed, according to these scientists, tells us that at the height of the Harappan civilization, the Ghaggar-Hakra had been flowing with enough water to support a large population.

However, about fifteen hundred years later, the physical evidence tells quite a different tale of this river. Geomorphologists say that, by 1000 BCE, the river system had essentially gone dry.[16] Indeed, archaeologists have found campsites set in the actual riverbed of the Ghaggar-Hakra that date to the period after the fall of the Harappan civilization. Consequently, river scientists theorize that the source waters of that critical stream system must have been diverted into another river system due in part to the region's frequent earthquakes. These scientists say that as the waters of the Ghaggar-Hakra were lost to the region, the Harappan cities and farms gradually became drier and drier.

That sounds reasonable. But those facts don't show any distinct climate change playing a significant role in influencing the actions of the Harappans. Is there any physical evidence that relates to climate?

As I mentioned earlier, there are two conflicting theories regarding the decline of the Harappan civilization. The first suggests that a prolonged drought led to its eventual deterioration while the second says that the drought was not a major contributing factor in the decline (and cites potential human causes such as the movement of Aryan peoples into the region).

Interestingly, the first theory, when two climatologists expressed it in the 1970s, also was linked to human causes. Those two Wisconsin climatologists, Reid Bryson and Thomas Murray, coined the term "human volcano" in reference to the potential impact of human-generated particulates on climate.[17] They suggested that widespread innovations associated with the farming practices of the Harappans eventually might have led to the society's gradual decline. Bryson and Murray reasoned, harkening back to the Dust Bowl days of the 1930s, that the dual processes of Harappan plowing and periodic drought might have sent a markedly increased dust load into the atmosphere and that created an irreversible climate change. They theorized that more dust in the atmosphere would have led to drier conditions. To combat those drier conditions, the Harappans would have put even more land into cultivation, Bryson and Murray reasoned, which in turn led to still more dust in the atmosphere. That would have produced, according to their theory, still drier conditions, and so on, until the land was completely transformed from farmland to desert. This is what scientists call a climate feedback response.

However, most scientists today discredit that line of reasoning. While cultivation processes can undoubtedly exacerbate a drought (as, indeed, we did witness during the Dust Bowl days of the 1930s or during the terrible Sahel Africa droughts of the 1990s), there simply isn't evidence to suggest that such practices can lead to irreversible changes in climate and environment.[18] Farming practices, we now believe, can make a drought worse but likely cannot create permanent environmental change—although some practices such as forest clearance may indeed lead to permanent change. So, while the concept is still considered viable of drought influencing the gradual decline of the Harappan civilization, the added influence of human cultivation practices on the underlying climate change of the region is now considered a relatively minor or even negligible factor.

Yet a good deal of the available physical evidence appears to support the basic underlying drought/social decline theory. One of the key indicators is the gradual change in the type of crops that the Harappans cultivated. Paleobotanical analyses of Harappan sites indicate an agriculture based primarily on wheat and barley. However, the paleobotanical evidence of late Harappan civilization—particularly by 2000 BCE—shows a much greater proportion of the fields were based on small millets and rices.[19] This shift in crop type suggests a gradual movement toward more drought-tolerant cereals and a greater reliance on two annual harvests (one associated with winter rains and the other linked to the summer monsoonal rains) and less dependence on presumably continuous water supplies such as rivers.

This somewhat subtle agricultural change may have led to even greater ramifications. For example, although cereal millets and rice are more drought tolerant, they, in general, tend to produce lower yields per acre than grains such as wheat and barley. In addition, the

Harappan economy depended on stored grains. Consequently, as the farmers began to shift toward millets, which do not store as easily as grains such as wheat, the region was less able to support (and maintain) a large urban population. More localized crops gradually replaced the human resources and the infrastructure to store and distribute grains.

So a climatologist might summarize the Harappan/environment interaction as follows:

- First, the Harappans took advantage of the wet environments and they became sedentary, or fixed in location, farmers in the Indus Valley region.
- Second, they then made major adaptations to their society to take best advantage of those environments. In essence, that meant they established villages and cities along major consistently flowing rivers.
- Third, the Harappans adapted as well as they could to slow environmental and climate changes. For instance, the evidence suggests that they shifted to more drought-tolerant, but less urban-supportive crops such as millets and rice as the climate shifted toward a drier regime.
- But, ultimately, the new environment of the Indus Valley (as seen in the loss of permanent river systems and the more frequent occurrence of drought) could likely not support the type of society that had developed there over the previous centuries, and so, over the years, the civilization declined.

So, let us go back to our initial two questions that we should ask before we conclude climate and environment caused the collapse of a society. First, did climate and environmental stability play a role in the initial formation of the culture? In terms of yes and no, I think we can answer that only partly. The physical evidence indicates that the initial wetter climate in the Indus Valley region may have helped encourage sedentary (rather than nomadic) agriculture, but periodic drought afterward may have helped consolidate the people into a river-based civilization where grain production was dependent on irrigation practices. Climate variability, coupled with consistent stream flow, likely influenced the society's initial growth but it did not "control" the Harappan civilization.

And our second question? Did climate change lead to the collapse of that society?

By 2000 BCE, a new, drier environment in the Indus Valley (with the loss of permanent river systems and the more frequent occurrence of drought) promoted such changes in the society that the new culture could likely not support the type of society of the previous centuries, and so, over the years, the civilization declined. Note that I said decline, not disappear. I think the evidence supports a fairly gradual change in population. Such a shift might be equated to the gradual movement of people from the Rust Belt states of the Northeast United States to the Sun Belt states of Florida, Texas, and the Southwest over the past quarter century.

The Harappans strike me as a people who first tried—fairly effectively—to use their environment and climate, then adapted as that environment changed. Eventually that adaptation, coupled with continued societal change, gradually led to the abandonment of the original urban civilization. Some might ask if we should label the Harappan civilization a failure since it did not succeed in adjusting adequately to the new environment of the region. I find that question intriguing since it implies that only if the preexisting culture had been maintained can the society be classified as a success.

So any change is a failure? I personally define a civilization's failure as the inability of the society to meet the current and near-future demands of its citizens. For example, society failed in my opinion, when, as Arab scholar Abd al-Latif recorded, a horrific drought and Nile flood in 1200 CE prompted the Egyptians of that time to undertake cannibalism and the actual bartering of babies for food.[20] In contrast, the Harappan civilization never appears to have experienced a massive die-off or sudden cataclysmic collapse. Their decline appears to have been relatively gradual and measured—no tumult or disaster—as the Harappans adapted to meet the increasingly harsher environmental and societal constraints imposed by and upon themselves.

Did the Harappans succeed in their civilization? I wouldn't presume to answer—I simply suggest that everyone develop his or her own measure of success to judge the Harappans . . . and us.

Time: Approximately 1900 BCE
Location: Near the Indus River, in modern-day northwest India, Asia

It had been another hard year. Déin shook his head ruefully as he leaned on his hoe. He surveyed the small millet field with its few scrawny grain heads sticking pathetically out of hard, dry soil along the bank of the failing stream. Yes, there would likely be enough food—barely—for his family to last over the long winter. But, if the monsoon rains didn't fall again next year, his family would be hard-pressed to stay here. There simply weren't those great surpluses of food that he remembered seeing as a young boy. And, if one was the slightest bit careless or had offended the gods, it seemed that death was always waiting to take that person away. Already this year, Déin had lost his youngest daughter to a strange, withering illness.

Of course, the village had already lost a great many of its youth over the past few years—many of the village's fewer and fewer young people were heading to presumably better lands to the east. His wife was already loudly commenting that there simply wasn't the number of potential women in the area to marry to his eldest boy, Jal, as there had been when Déin was a boy.

And that was true. Life today surely wasn't like the stories of the great and wondrous cities that Déin's grandfather had loved to tell him about when he was a small child.

Déin glanced back toward his thatched hut. Only last night his wife had said that it was utter foolishness to force his eldest boy to stay here by this slowly dying rivulet of a stream. She had argued that he should let Jal leave as the boy wished so that Jal could try his fortunes in the wetter lands to the east. Perhaps she was right, Déin thought as he glanced up at the cloudless blue skies that failed to even hint at a promising shower. Perhaps she was right.

Chapter 6

THE MYSTERY OF THE EXODUS

Religion and weather have often been tightly intertwined in the long history of civilization. Indeed, many of the gods of early societies were directly linked to specific aspects of nature—in particular, storms, lightning, and rain. For example, the favorite weapon of Zeus the Stormbringer was the lightning bolt. His Roman counterpart, Jove, also was considered a master of lightning. In a similar fashion in a far colder place, the Norse god Thor was deified as a powerful god of weather. Lightning was the result of sparks streaming from his magic hammer, Mjollner, as he hurled it to Earth. Even David in the book of Psalms calls upon Jehovah to "*Cast forth thy lightning and scatter them; shoot out thine arrows and destroy them.*"[1]

Consequently, we often find weather and religion intermingled throughout the histories and legends of many early peoples. In terms of the Western religions, one sequence of profound religious events common to Judaism, Christianity, and Islam has recently been interpreted in the light of modern meteorological knowledge. With the idea that religious events might have their origins in actual meteorological phenomena, let us take a look at the great weather mysteries of the Exodus.

Time: March 29, 1281 BCE
Location: Near the Red Sea separating Egypt and the Sinai, Africa

Simon was becoming increasingly worried; he, his young family, indeed all of the increasingly desperate people with him, seemed to be caught between a rock and a hard place. Ahead of them stretched an immense body of water—what his friend Joseph had overhead the Elders call "yam suph," or "the Sea of Reeds." A few of the more adventurous and able of the One God's People, specifically the advance scouts, had traveled ahead of them throughout the long journey. They had already tested the waters and found them relatively shallow—seldom above chest high. Still a growing despair was enveloping the struggling band of refugees. It was quite obvious that the whole People could not pass through the placid waters of the Sea of Reeds—not laden as they were with small children, their animal herds, and all of their possessions.

And now word had come up from the rear of the long caravan that some had seen the Pharaoh's army marching in the distance behind them. Rumors were spreading through their camp that the Pharaoh had reconsidered his decision to let the One God's People leave and had ordered his warriors to recapture them.

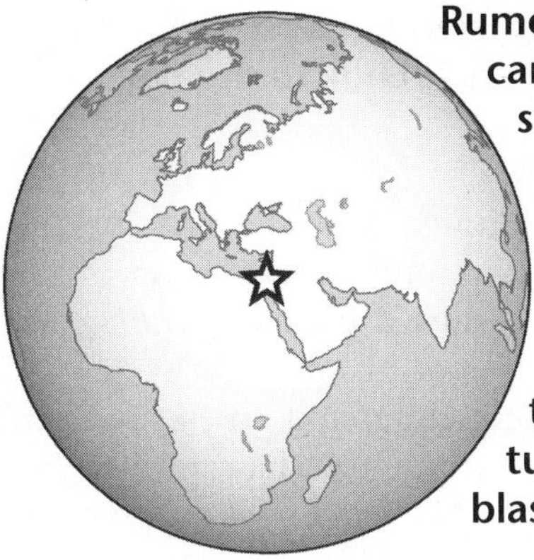

His wife Sarah, with their two small babies nursing on her chest, turned to him with concerned features on her angular face. The blasting easterly windstorm that had

sprung up over the night whipped some strands of her brown hair from the cover of her scarf.

"Dear husband, what will become of us?"

The journey had come to an abrupt end after the many wearying days of traveling around the wilderness to avoid the more populous centers. The Elders, speaking for Aaron and his holy brother, said that they would first travel south, thereby avoiding the land of the Philistines. Simon's knowledgeable friend Joseph had whispered that this would be to keep the People together without contamination by unbelievers and to limit potential communication of the People's movements to the Pharaoh. Simon didn't have an opinion about that but, whatever the reason, the decision to avoid the populated Mediterranean coast had led them to this desolate place. Now, he thought sourly, their way to the east to the promised land of milk and honey was blocked by the stagnant waters of the Sea of Reeds. And worse, according to Joseph, who always seemed to have the most up-to-date information, troops from the Lord of the Two Lands might be marching behind them to catch them, as they stood stranded here on the western shore of this stinking sea.

So what would they do? He glanced around at the litter-strewn area with worry. On top of everything else, this horrendous windstorm that had sprung up last night was wrecking their camp.

"Simon!"

He looked up as his good friend Joseph trotted over to him. "Simon! Look to the east! Look to the east! You won't believe it!"

According to the biblical book of Exodus, after the tenth plague of the death of the Egyptian firstborn and Yahweh's Passover of the Israelites, the Pharaoh said to Moses and his people, *"Take your flocks and your herds, as you have said, and be gone."*[2]

The Israelites followed the Pharaoh's command but, as written in Exodus, took a rather roundabout journey from Egypt following Yahweh's admonishment, *"Lest the people repent when they see war, and return to Egypt."*[3] In a military sense, this was excellent advice since, at this time, the Peoples of the Sea, a loosely governed collection of seafaring raiders, were invading the northeast coastal areas of Egypt.[4] Had the Israelites taken the straightforward coastal route, they undoubtedly would have encountered—and had to fight—these formidable invaders.

Consequently, the Israelites took a longer route *through the way of the wilderness of the Red Sea.*[5] The Red Sea is a large, water-filled rift basin separating Africa from Asia. It averages only about 175 miles or so in width but extends for nearly 1,500 miles toward the Indian Ocean. Climatically in the modern world, it is quite inhospitable—geographically, it separates two very arid deserts, the Arabian and the eastern Sahara. While the region receives almost no rainfall, its air is uniformly humid and hot.

As they traveled through this harsh environment, Yahweh gave the struggling Israelites an astonishing physical manifestation of his presence: "*And the Lord went before them by day in a pillar of dust to lead them along the way.*"[6] This concept of God manifesting as a weather phenomenon such as a whirlwind or tornado is also found later in the Bible. For example, the great Israeli prophet Elijah was lifted up bodily by a whirlwind to heaven at the Jordan River. Specifically, in 2 Kings, it is written, "*And Elijah went up by a whirlwind into heaven.*"[7]

In terms of climate, dust devils, or small vortices of wind, sand, and dust, are common in the desert. They are created by intense surface heating that causes the air above it to rise. Although related to tornadoes in that both tornadoes and dust devils are spinning vortices, a tornado is created by the wind processes of major storms while dust devils are formed from the air currents heated by hot desert surfaces. If local wind shear (a vertical change in wind speed and direction) can cause that rising air to spin, a dust devil will sometimes form. Dust devils are generally weak in wind strength with peak speeds averaging thirty to forty miles per hour. Normally, dust devils are rather transitive weather events with most lasting only a few minutes. A scientist in 1947 observed what maybe the longest-lasting dust devil in modern times.[8] He watched a single dust devil spin on the Bonneville Salt Flats in western Utah for a remarkable seven hours. During that time, he reported that it grew to twenty-five hundred feet in height and traveled for forty miles.

Aerial view of the desert of eastern Egypt, a region likely crossed by the ancient Israelites.

In Egypt, the land of the Exodus itself, a French naturalist in 1873 observed a dust whirl two miles northeast of Cairo that began at 10:30 in the morning on a small sand mound, remained stationary for nearly an hour, then in response to a gentle breeze, wandered away.[9] But it nevertheless maintained its sharply defined outlines and great altitudes for more than three hours until it was lost in the distance, lasting for a total of at least four and a half hours. At noon the naturalist estimated the dust devil's height at over three thousand feet high and said it was completely opaque.

So was Yahweh's *pillar of dust* a simple dust devil? Perhaps, but I believe that a much more plausible physical explanation for the Exodus "pillar of dust" might be formulated when its nocturnal counterpart is also considered: "*And the Lord went before them by day in a pillar of cloud to lead them along the way, and by night in pillar of fire to give them light.*"[10] What that means is that the Israelites traveled in the direction of a dust column in the day and a fire column by night. The combination of two distinctly different types of manifestations—a column of smoke and a pillar of fire—suggests a very interesting possible physical explanation.

In his book *The Tornado*, J. E. Weems made the fascinating revelation that Apache Indians in the Desert Southwest of North America would sometimes create their own dust devils for use in desert signaling by setting a cactus on fire and waiting for the swirling updraft to suck up dust.[11] What if the Israelites did exactly the same thing by sending out advance scouts ahead of the main group of refugees to pick the best route? Rather than travel all the way back to the main group, wouldn't it be more efficient for these Israelite advance scouts to transmit their message of the safest travel path by using the signaling method employed three thousand years later by the Apaches? Couldn't the advance scouts burn desert scrub during the day, thereby creating a dust column? Although there aren't the tall saguaro cacti of Arizona in that part of the world, there are other types of vegetation. And if the Israelites employed the same technique at night, wouldn't the main party see the fire pillar of that vegetation burning?

But, according to the Bible, the situation with the column of smoke abruptly changed when the Pharaoh's army was sighted advancing behind the Israelites: "*The angel of God who went before the host of Israel moved and went behind them; and the pillar of cloud moved from before them and stood behind them coming between the host of Egypt and the host of Israel . . . and the night passed without one coming near the other all night."*[12] If we follow the Israelite logic of sending advance scouts to light smoke signals ahead of the main party in order to guide the wandering Israelites, it is plausible to extend that line of thought to this new occurrence.

Signal smokes and fires can be seen by not only those whom the signalers want to see such signs, but by anybody else in the immediate area. As the Egyptians advanced on the stalled Israelites at the Sea of Reeds, it is logical to speculate that the Israelite leaders realized their precarious position. Consequently, it is reasonable to suggest that once the Pharaoh's army was sighted, the Israelite leaders ordered those advance scouts to quickly backtrack to the main party and then journey west into the Egyptian desert to light new signal fires—thereby hopefully misdirecting the Pharaoh's army from where the main party was camped!

Beyond this possible militaristic interpretation of the smoke/fire manifestations, the critical event of this portion of the Exodus was set to occur over the night as the smoke obscured the Israelites' position from the advancing Egyptian army:

> Then Moses stretched out his hand over the sea; and the Lord drove the sea back by a strong east wind all night, and made the sea dry land, and the waters were divided. And the people of Israel went into the midst of the sea on dry ground, the waters being a wall to them on their right hand and on their left. The Egyptians pursued, and went in after them into the midst of the sea, all Pharaoh's horses, his chariots, and his horsemen. . . . Then the Lord said to Moses, "Stretch out your hand over the sea, that the water may come back upon the Egyptians, upon their chariots, and upon their horsemen." So Moses stretched forth his hand over the sea, and the sea returned to its wonted flow when the morning appeared; and the Egyptians fled into it, and the Lord routed the Egyptians in the midst of the sea. The waters returned and covered the chariots and the horsemen and all the host of Pharaoh that had followed them into the sea . . . but the people of Israel walked on dry ground through the sea, the waters being a wall to them on their right hand and on their left hand.[13]

The waters of the Sea of Reeds parted, allowing the Israelites to pass across the previously flooded area. Can there be a physical explanation for such a seemingly miraculous event?

One possible explanation that has been offered is the occurrence of a tsunami. What makes this an attractive theory is that the region is geologically active since the area forms the boundary between continental plates, specifically those of Asia and Africa. As I mentioned in the Mystery of the Dead T. Rex, a tsunami, or as it is sometimes inappropriately termed, a tidal wave, is generated by volcanic or seismic activity, such as an earthquake, or by an asteroid impact. The term tsunami was adopted in 1963 by an international scientific conference as the correct term for a series of waves generated in a body of water by a disturbance that suddenly displaces that water. Tsunami is a Japanese word represented by two characters: *tsu* and *nami*. The character *tsu* means harbor, while the character *nami* means wave.

As a tsunami crosses a deep ocean, its length from the crest of one wave to the crest of the next may be a hundred miles or more, and its height from crest to trough will only be a few feet or less—but the waves will reach speeds exceeding six hundred miles per hour. Because of their small vertical size in the deep ocean, they cannot be felt aboard ships nor can they be seen from the air. However, when the tsunami enters the shallower waters of the coast, the speed of the waves rapidly diminishes—and the wave height dramatically increases. In some shallow coastal waters, a large tsunami might reach a crest exceeding a hundred feet! The primary determinant to the size of a tsunami is the initial vertical displacement of the land under which the earthquake occurs. However, as the tsunami propagates across water, many other factors come into play, including such geologic considerations as the shoreline configuration, the slope of the ground from water to land, and the water depth.

Researchers have determined that, for the Exodus, if an earthquake at the southern part of the Red Sea produced an initial wave of, say, three feet, the tsunami would have grown to nearly ten feet by the time it reached the Gulf of Suez. A key point is that a wave has a fifty-fifty chance of entering into a region at its crest or at its trough. This means that half of the time, the tsunami will strike as a massive surge of water—such as the deadly December 26, 2004, tsunami that smashed the coasts of the Indian Ocean. But a tsunami also has the same probability of coming ashore at low water as a trough with the water suddenly receding from the shore. This was the case in the 1963 earthquake-generated tsunami that struck the

Hawaiian Islands. Tourists on the beach suddenly were witness to the sea sucking itself back, exposing a mile or more of previously submerged coast. Some of these tourists, not realizing their peril, journeyed out into that suddenly dry land to collect shells—when the crest of the wave rose from the sea and swept back over the area.

Some theorize that the Exodus crossing can be physically explained by such an event. H. Goedicke attempted to account for the parting of the Red Sea by theorizing that a volcanic-induced tsunami followed the eruption of the volcano of Santorini in the Aegean Sea.[14] He postulated that the lowering water level of the Sea of Reeds was the tsunami coming at trough, rather than crest. The collapse of the waters on Pharaoh's army following the Israelites' safe passage was the arrival of the tsunami's crest.

Researchers Doron Nof and Nathan Paldor, two brilliant geophysicists investigating physical possibilities to explain the Exodus crossing, pointed out a couple of shortcomings with a tsunami theory.[15] First, they noted that the biblical account states that the recession of water was the result of a long night's strong easterly wind—something that would not be created by a tsunami. Second, the time frames, as described in the Bible, for the recession and the inundation of waters were not of equal length—the recession took place over a longer period of time, perhaps as much as an entire night, while the inundation took place abruptly.

Consequently, they suggested another physical explanation of the Exodus crossing; specifically the idea of *wind setdown.* They examined the scientific possibility that the waters of the Sea of Reeds could have receded as a result of a prolonged (the Bible states "overnight") wind that blew along the Gulf of Suez. The two researchers determined that a storm of even moderate strength (with winds of at least forty miles per hour) could move the water such that sea level in some areas could drop by as much of eight feet. Such a large drop, they reasoned, would be primarily the result of the two factors: the particular geography and alignment of the Gulf of Suez—being a long, narrow, and relatively shallow body of water—and the complex nature of the fundamental physical relationship that governs the forces creating our world's winds.

Nof and Paldor noted, for example, that the waters of Lake Erie—which is somewhat similar in shape to the Gulf of Suez—often fluctuate by as much as three feet depending on the strength and direction of the winds. However, because Lake Erie's slope from shore to deep water is much steeper than that of the Gulf of Suez, no significant water recession (appearance of land) is evident for that lake.[16]

Of course, the recession of the waters—allowing for the passage of the Israelites—is only half of the Red Sea miracle. Does the theory of these two researchers account for the subsequent reflooding of the land—and the death of the Pharaoh's army sent to recapture the Israelites?

The key is the speed of the wave release. In other words, if the winds could push up the water on one side of the Sea of Reeds such that bare ground was exposed, how fast would the waters reclaim that ground when the winds subsided?

Normally winds gradually relax over a period of hours to days. Such a gradual subsidence would allow the waters to slowly shift back to their "prewind" state. If, however, the winds suddenly subsided within the space of, say, a few minutes, then the water that had been pulled by the winds would rebound back as what is termed a *gravity wave*, or a *bore*. Scientifically, a bore is described as a sudden elevation of water in the form of a hydraulic jump. It is observed just at the instant that the river flow changes from ebb to flood and occurs at the time of especially large tides. A tidal bore is viewed as a vertical wave that moves upstream. But Nof and Paldor specifically said that tides were not necessary for the formation of a potential Red Sea bore. Their calculations indicated, given a return speed of roughly ten miles per hour, that the entire area, which had been emptied by the prolonged winds, could be flooded in just four minutes—without consideration of tides.

An Israeli researcher, M. Dayan, originally proposed the possibility of tides aiding the Exodus.[17] Nof and Paldor stated, "Although the wind-tide combination is certainly possible, we find it hard to believe that [normal biweekly] tides were *crucial* [italics from original paper] to the crossing. This is not only because of the extremely low probability of [a biweekly tidal] mechanism, but also because we speculate that the Egyptians . . . could not have been ignorant of a tidal regime that repeats itself every two weeks. Being aware of such a tidal regime, they could easily avoid getting trapped on a temporarily exposed ridge."[18]

And, in my opinion, Nof and Paldor are correct in assuming that many ancient civilizations were aware of *normal* tidal activity in the area. In fact, the first known historical writings concerning tides (by the ancient Greek scholar Herodotus as cited by historian George Rawlinson) mentioned the daily "ebb and flow of the tide" occurring in the Red Sea and Gulf of Suez.[19] Other ancients also mentioned the tides of this area. For example, the ancient Greek geographer Strabo reported on the views held by Posidonius (51 BCE) and Aristotle on the tides near Cadiz and by Seleucus on those in the Red Sea.[20]

But what if there was a combination of a sustained strong wind event, such as identified by Nof and Paldor, *and* a rare tidal situation? Could that provide a more complete scientific explanation of the biblical Red Sea crossing?

The first item that we would need to determine is the range of the normal tides for this region. The Defense Mapping Agency (DMA) of the United States suggests that the normal tides in the Gulf of Suez average from forty centimeters (about sixteen inches) in the south central Suez, but according to the DMA, "with strong southerly winds, which are prevalent during the winter months, the tide may rise as much as 8 or 9 feet in the bay [of Suez, located in the extreme northern portion of the Gulf of Suez]."[21]

Aha! That would suggest a combination of both winds and tides might produce major changes in sea level for parts of the region. But is there any way to compute tides far back to the time of Moses and the Exodus?

Beyond the normal biweekly variability, tidal heights can also vary dramatically from year to year, decade to decade, or even century to century. Fortunately, a gifted former student of mine, Dr. John Shaffer, and I demonstrated that such long-term variations in tides could be mathematically computed using the concept of equilibrium tides.[22] Global, or equilibrium, tidal variations are a function of short-term lunar and solar astronomical variations, as well as the geologic timescale variations involving the Earth-moon orbit—which are known and can be calculated.

So, if one accepts the reasonable assertion by Nof and Paldor that the Egyptians would likely have known the *normal* biweekly tidal regime, the tidal activity potentially *unanticipated* by the Exodus's Egyptians would likely be a more extreme and rarer event. A possible candidate is the phenomenon of spring tides. Spring tides occur under identifiable orbital configurations of the Earth, moon, and sun, specifically when the moon is in close proximity to the

Earth. The phrase "spring tide" is not indicative of a particular seasonality; rather the term is defined as the specific tide when the ranges between high water and low water are greatest. Indeed, currently, the greatest spring tides do not occur in the Northern Hemispheric spring but rather during the September equinox.

The magnitude of extreme spring tides is a matter of historical record. For example, a spring tide event that occurred during the fall 1935 equinox (one of the two times of the year when the direct rays of the sun are over the equator), computed via our tidal model to be the largest of this past century, produced the following *London Times'* comment: "The biggest tides in the memory of river men have swept up the River Trent at Gainsborough. Before the eagre, or tidal bore as it called on the Severn, came this morning it was possible to walk across the river at certain points and shipping was held up."[23] Given such variability in water level, an extreme spring tide emerges as a possible candidate for an unanticipated tidal event during the Exodus crossing.

If a spring tide event coupled with strong winds is a potential candidate for a rare tidal event, we can begin to fine-tune a time for the Exodus. Although currently the largest spring tides occur in September, the greatest spring tides for the globe prior to 950 CE shifted to the time of the March equinox, due to variations in Earth's orbit. Consequently, if extreme spring tides did influence the Israelites' Red Sea crossing, the most favorable seasonal time frame for the Exodus crossing, occurring as it did before 950 CE, would have been during spring rather than in autumn.

For the Levant region, spring is also the season in which the strongest winds are most likely to occur. Modern observations taken at the stations along the Gulf of Suez indicate that, in general, April has more storm events—and stronger winds—than any other month of the year. Indeed, because of that information, Nof and Paldor used April as the exemplary month for their wind setdown theory. Consequently, both tidal and wind analyses favor the selection of a spring month as the most likely time for the Exodus crossing to occur. Is there collaborative historical evidence for such a spring date?

Jewish Passover is traditionally celebrated on the sixteenth of Nisan, which should be the first spring month.[24] In pre-Babylonian Exile times, the name of the month in which the Exodus took place, and in which the Passover was celebrated, was Abib, as noted in

Exodus, "*This day came ye out in the month Abib.*"[25] Abib may be translated as "kernel of grain" and marks the time of the year when such kernels appeared. During the Babylonian Exile, however, the Jews made use of the Babylonian calendar and they kept that throughout their later history, down to the present day. Thus the month earlier known as Abid, in which Passover and the Exodus were celebrated, became Nisan as mentioned in Nehemiah, "*And it came to pass in the month of . . .*"[26] In all cases, it appears that the Jewish Passover (and the Exodus) is celebrated as a spring event. Although it is possible that such a date also served as an agricultural festival long antedating the time of Moses, such a spring date is compatible with the most likely dual occurrence of strong winds and extreme high tide.

The traditional celebration date of Passover, following the Jewish lunar calendar, would be the time of the full moon. Tidal, wind, and historical evidence for the Exodus, therefore, suggest a seasonal time frame of spring (with April the most probable month) during the time of a full moon, given the Jewish calendar.

Fundamentally, however, the question of the *necessity* of spring tides to an Exodus crossing solution still remains. That question may be addressed by calculating the tidal ranges determined from the tidal model. The modern (1700–2000 CE) average daily equilibrium tidal range for the globe is 0.72 meters as computed using Shaffer's and my astronomical tidal model. The corresponding global average spring tidal range is 0.84 meters. The greatest equilibrium tides computed over the last three centuries occurred during the autumnal equinoxes of 1829 (0.92 meters) and 1935 (0.92 meters). Both of these extreme springs averaged approximately 27.4 percent above the long-term mean height of tides.

Scholars traditionally place the time of the Exodus at around 1250 BCE. Our tidal computer model suggests three possible extreme spring tides around that time linked to the Exodus that were markedly larger than those of today. The largest tidal range (1281 BCE) is computed to be 0.92 meters, or 27.4 percent above the long-term present-day mean. Between the years 1300 BCE and 1200 BCE, only three days (or 0.008 percent of the total days within the hundred-year period) meet the criteria of (a) an extremely high equilibrium tide (greater than 0.926 meters) and (b) an occurrence near the full moon (the middle of the month of Nisan). Those three days are (a) March 31, 1281 BCE, (b) April 3, 1246 BCE, and (c) April 5, 1219 BCE.

(a) *March 31, 1281 BCE.* Conventional chronologies generally place this date within the reign of Pharaoh Seti I. The tide associated with this astronomical alignment was the largest experienced within the hundred years from 1200 to 1300 BCE and indeed is matched by only one other tide within 130 years prior to that hundred-year envelope. The seasonal time frame for the event is the earliest of the three extreme tides within the envelope. The event occurs after the spring equinox and near the start of the year's stormiest month of April. The event was associated with a full moon.

(b) *April 3, 1246 BCE.* Conventional chronologies generally place this date in the middle of the reign of Pharaoh Ramses II. The extreme tide of this event was the third largest in the hundred-year envelope around 1250 BCE and it followed the 1281 BCE extreme tide by thirty-five years. The date of this extreme event places it within the storm activity common for the month of April. This event occurred near the date of a full moon.

(c) *April 5, 1219 BCE.* Conventional chronologies generally place this date near the end of Pharaoh Ramses II's reign. The extreme tide of this event was the second largest of the hundred-year envelope and followed the 1246 BCE event by twenty-seven years and the 1281 BCE extreme tide by sixty-two years. Seasonally, this event is the latest of the three events and is within the time of maximum storm activity. As with the other two events, this event occurred near the date of a full moon.

Of these three dates, geophysical considerations alone would rank the 1281 BCE event as the most favorable, followed by the 1219 BCE event, and the 1246 BCE date as the third most favorable. However, the magnitudes of all three events are very similar and any of the three, in concert with sustained winds or alone, could have produced sea level reductions on the order of ten to eighteen feet (3.0 to 5.5 meters).

Nof and Paldor gave probabilities for a ten- to fourteen-hour sustained wind event of 20 meters per second (about 40 mph) as once every thousand to three thousand years. A corresponding probability for the occurrence of tides equal or greater than the three discussed above for a five-hundred-year period around the Exodus crossing is approximately one event per seventy-five years. For the maximum drawdown, or lowering, of water (5.5 meters, or roughly 18 feet), the probability of a coupled occurrence of the two phenomena (extreme tides and sustained winds) is defined as the product of the two mutually independent probabilities. Consequently, the combined probability of an extreme tide and a sustained wind event is approximately once every 150,000 years, making such a combination of sufficient rarity, I believe, to almost warrant the term "miraculous."

It must be emphasized that this investigation examines only the *most favorable possibilities* for a calendar date of the Exodus. Because neither the specific geography nor the environment of the biblical Red Sea crossing—and, consequently, the sea level reduction needed for a crossing—are known, a number of equally possible geophysical alternatives must also be considered. For example, as I discussed above, Nof and Paldor created a scenario involving a sequence of windstorms that, without addressing tides, is a viable explanation (if the hypothesized ridge was submerged less than eight feet). Or, alternatively, weaker tidal events than those described above, when coupled to the Nof and Paldor wind setdown model, may also adequately address a Red Sea crossing, given a more moderate depth for the presumed submerged ridge.

So the possibility of an actual physical explanation for the Red Sea crossing of the Exodus exists. Does this mean that God did not personally intervene to save the Israelites? Perhaps Doron Nof and Nathan Paldor phrased an answer best when they wrote, "Believers can find the presence and existence of God in the creation of the wind with its particular properties just as they find it in the establishment of a miracle."[27]

Time: March 29, 1281 BCE
Location: Near the Red Sea separating Egypt and the Sinai, Asia

"Hurry, Simon," his beloved wife Sarah called back to him.

Simon had paused for a moment, gazing at the wondrous sight: the low-lying swamp had actually retreated! It had withdrawn back perhaps an amazing hundred paces or more! A long strip of wet and

marshy—but quite passable—land now lay before them and the rest of the People. The astounding strip of marshland extended across the Sea of Reeds to the sandy hills of the east.

"Hurry, my husband," Sarah again called out, as she turned into the blasting wind still roaring from the east. The elders, led by Aaron and his holy brother, had now descended to what had yesterday been the shore of the swampy sea and were beginning the trek across the miraculous sliver of land that extended before them.

"How can this be?" Simon gasped as he jogged up to his wife and took one of the sobbing twins from her arms.

"It is the will of the One God," his wife replied with the absolute certainty of faith. "He has blessed His People with this miracle!"

Simon shook his head in wonderment. He was just a common man—not like the great elders such as Aaron or his holy brother—and this strange sight below completely befuddled him. How could such a bizarre event occur?

But perhaps, he suddenly realized, it didn't matter how this incredible oddity had actually come to pass. Perhaps the most important thing was that the People would survive after all—they would safely pass into the east, away from the Lord of the Two Lands. And that meant that, just perhaps, Simon's People were indeed destined for greatness as the elders claimed. He glanced down at the tiny baby in his arms.

"Perhaps, little one," he spoke over the continued howl of the windstorm. "Perhaps one day our People will indeed have a land that we can call our own."

Chapter 7

THE MYSTERIES OF WEATHER AND CLIMATE IN ANCIENT GREECE

Homer's majestic epics involving the fall of Troy and the subsequent wanderings of Odysseus are set in a time in ancient Greece that we now know actually existed. Heinrich Schliemann, a nineteenth-century archaeologist (or potentially fraudulent treasure hunter, depending on one's viewpoint), was one of the first to discover the ruins of the civilization that has been linked to the fall of Troy, as described by the great, supposedly blind Greek poet Homer. He worked under the assumption that the stories told by Homer were based on actual events. He accordingly interpreted the ruins of Mycenae in southern Greece using the *Iliad* and the *Odyssey* as guidebooks. Most dramatically, as the scholar Jennifer March reported, when Schliemann uncovered a golden death mask resting upon a skeleton in one of the tombs, he purportedly wired the British prime minister with the message "Today I have gazed upon the face of Agamemnon."[1]

Following Schliemann's discoveries, many more excavations (undertaken in a decisively more scientific fashion) were conducted both in Mycenae and in sites associated with ancient Troy. In fact, recently, a team of scientists led by geologist John Kraft presented results of their comparisons of the modern geology of the region with the landscapes and coastal features described in the *Iliad* and other classical sources.[2] They concluded that there are many marked similarities between the supposed location of Troy as identified by Schliemann and descriptions of the landscape given in the *Iliad*.

Based on their excavations, modern archaeologists now suggest that the layer of ruins at the site at Troy known as "Troy VIIa" is the most likely candidate for the Troy so elegantly described in Homer's *Iliad*. These ruins date back to the mid- to late thirteenth century BCE. Although early excavations of the last century suggested that this site was only a small hilltop fort, more recent excavations in the late 1980s indicate a city of perhaps considerable size. Most important, archaeologists discovered evidence that Troy VIIa was destroyed by war and fire. Consequently, modern archaeology strongly suggests that Homer apparently was describing actual events.

But how did climate and weather play a role in Homer's ancient Greece?

Time: Approximately 1260 BCE
Location: The coast of the Aegean Sea, within a few miles of the ancient city of Mycenae, Greece, Europe

"Grandfather, tell me again of the Great War of Troy."

The bright-eyed boy with the straw-colored mop of dirty hair was perhaps seven or eight years old. He sat with an attentive gaze at the feet of the grizzled, white-bearded man. The two rested outside the family's villa-style house overlooking the beautiful Saronic Gulf in the early morning cool breeze. The brilliant blue cloud-

less sky foretold that today—as had many of the days of this spring—would once again be dry and warm. In the distance, the old man could see the family's servants already working the estate's grapevines—tying off branches and pruning the plants—from which the family produced its regionally famous wines.

The old man smiled fondly at the boy. The youngster's request was a familiar and frequent one. The boy had the soul of a warrior; he constantly yearned to hear of the old days of war and daring-do rather than be forced to live the mild life of the grandson of a country landowner.

"Well, young Hythamer, I was but a young man back then, perhaps only ten summers older than you are now. Ah, how anxious we were to fight for our exalted King Agamemnon." The old man's gaze turned inward, remembering what, to his impressionable younger self so many decades ago, had been a truly awe-inspiring sight. Ship after deadly Greek warship lay peacefully at anchor in the harbor of Mycenae, each preparing to carry its troops of warriors across the water to glorious battle.

"The great king," the white-haired soldier said, settling into his storytelling voice, "had ordered his fellow kings from across all of the land to create the greatest naval fleet the world has ever seen—a thousand of the finest black-hulled ships—to set sail across the Great Sea to fight the evil horse-worshippers . . ."

"Master?" The hesitant voice of the head servant interrupted the old man's epic narrative.

The old man refocused his attention back on the present.

"Yes, what is it, Bellus?" he asked, irritable at the interruption.

"The estate's well has continued to drop very low, Master," the slave spoke with a deep worry underlying his words. "I am very concerned that if we should continue to use it to irrigate the grapes, we won't have enough for the household."

The old man nodded with a grimace. In contrast to the glories of the Trojan conquest, now the almighty gods appeared to have withdrawn their favors to the glorious kingdom of Mycenae over the last few decades. Each year seemed to reintroduce the haunting specter of drought to the parched land. It was one thing to win a decade-long war against a powerful enemy, but it was quite another to be able to run a large estate for a decade in seemingly constant aridity. Twice this spring already, he had been forced to order his servants to trek down to the river and then to laboriously haul the heavy water-filled amphorae up the steep pathway to his hillside estate and its parched grapes.

The old man silently considered the situation for a long moment, and then sighed.

"All right, Bellus. Take four of the strongest men and march them to the river. Try to be back by high sun."

The chief servant bowed his agreement and headed out to the servants' quarters.

The old Greek landowner leaned back in his chair and looked down at his young grandson.

"Now where were we, young warrior? Ah, yes, I was telling you that the great king had assembled the greatest fleet known to man—a thousand black-hulled ships filled with only the most heroic of Achaean warriors—to sail . . ."

With ancient Greece, we might examine both a weather mystery and a climate puzzle. First one might ask whether there is any proof of the validity of Homer's epics, the *Iliad* and the *Odyssey*, that one might gain by studying the weather. As mentioned above, archaeology has given us some tantalizing indications that a major siege and destruction of the ancient city of Troy did occur at a time that would be consistent with the ancient Greek civilization mentioned in Homer's great works. But can meteorology shed any new light on the possible reality of Homer's epics?

Second, following the successful siege of Troy and its subsequent destruction, ancient Greece—and specifically the region of Mycenae controlled by King Agamemnon—declined for the next five hundred or so years into an ancient dark age. Only after that dark age did the great classical Greek civilizations of Athens and Sparta rise to prominence. So our second mystery might be to investigate whether climate played a significant role in the decline of that first great Greek civilization.

These two questions point out a fundamental difference between climate and weather. To answer the first question—can meteorology shed light on the authenticity of Homer's *Odyssey* and *Iliad*—we must see if we can actually tease *daily* weather information out of these ancient epics. For the second puzzle—did climate influence the long decline of Mycenaean civilization into a five-hundred-year dark age—we must employ some interesting long-term climate identification techniques.

THE MYSTERY OF HOMER'S *ODYSSEY*

A few years back, I undertook an unusual study to see how much weather information we might extract from the great poet Homer's *Odyssey*, an epic saga following the Mediterranean travels of the celebrated Greek warrior Odysseus and his crew after the fall of Troy.[3] But before I review that study, three important factors about using this type of "historical" document must be addressed. First, the epics of Homer were not written at the supposed time of Troy's fall (in the thirteenth century BCE), but possibly as much as five hundred years later. Consequently, even if the assumption is made that some events described in the *Iliad* and parts of the *Odyssey* were real, error—and, undoubtedly, storytelling license, such as I demonstrate in my little vignettes for each of these mysteries—may have infused more drama into these poems. Second, these great epics were obviously not designed as meteorological logbooks or even military histories, but rather as great adventure and moral stories. For example, when I attempted (as you will see below) to trace portions of the Greek fleets across the Aegean Sea from Troy to the Greek mainland, I had to sort through many different parts of the *Odyssey* to do so. The great epic was not written in a straightforward chronological style. And the third caveat I needed to recognize is that weather wasn't considered a natural phenomenon by the people of these ancient times. Winds and storms were not the result of physical forces such as *pressure gradient forces* and *convection*, but were, rather, the direct weapons or rewards of the gods' personal anger or favor—and were specifically directed to individuals based on their interactions with their gods.

So, with those cautions, can we derive any useful weather information out of the *Odyssey*?

The *Odyssey* begins at the end of the Trojan War when the Greek fleet was about to set sail to return to the Greek mainland. What happened those first few days after various elements of the Achaean fleet began the crossing is best told by three commanders—Nestor, Menelaus, and Odysseus. Additionally, mention is made of three other Achaean leaders. In book III, Nestor recounted the fate of Diomedes of Argos, and in book IV Menelaus related, via Proteus of Egypt, the tragedies of Ajax and King Agamemnon.

On Day One from Troy, as told by King Nestor of Pylos, half of the Greek fleet stayed at Troy with Agamemnon to conduct a set of ceremonies to appease the goddess Athena. Meanwhile, following an argument, the remainder, including Nestor, Menelaus, Odysseus, and Diomedes, launched their ships at daybreak and landed at an island near Troy called Tenedos. There a debate ensued as to whether to return to Troy (and Agamemnon) or to continue to Greece. Again, the fleet divided: Nestor and Diomedes sailed on to Lesbos while the "wise and crafty" Odysseus made his way to Troy. Nestor and Diomedes made an uneventful landfall on Lesbos ("A clement god had smoothed the sea's rough surface").[4] Meanwhile, Odysseus returned to Troy. Ostensibly, the Greek hero's return was to help Agamemnon with the appeasement rites. But, from a weather viewpoint, a more pragmatic reason for the return of the undoubtedly "wise and crafty" Odysseus may relate to the generally weak winds observed by King Nestor.[5] A southerly route across the Aegean via Lesbos would not be favored unless northerly or northeasterly winds were anticipated. As a crafty sailor, Odysseus might have envisioned a faster—and safer—voyage than that of his fellow Greeks by backtracking north and then traveling westward along the Thracian coast. This would be sound reasoning if the weak winds of Day One were to persist. Odysseus's subsequent travel to Ismarus on the Thracian coast supports this argument.

But there are indications that, by Day Two from Troy, the winds began to pick up. Homer noted that Menelaus overtook Nestor and

Diomedes on the island of Lesbos—although Menelaus started later than his two fellow Greek kings in leaving Tenedos. Odysseus said that "the wind that carried me from Troy brought me to Ismarus."[6] Such a wind would have to be easterly or southeasterly, given Ismarus's location on the Thracian coast.

Meanwhile, the three Greek commanders on Lesbos were debating the best route to use in crossing the Aegean. Two different routes were proposed. The first course was shorter, but more dangerous—sailing directly across the Aegean to Geraestus, the most southerly point of the Euboea, just off the Greek mainland. The second route was longer but safer—it consisted of traveling first east and south to the island of Chios and then across the Aegean via island-hopping through the Cyclades. The ancient Greeks tended to fear long sea voyages away from sight of land, given the limited geographic knowledge of the sailors of the time (indeed that is one of the moral lessons of the *Odyssey*: be careful when you stray from home).

So, given this fear of long open-sea voyages, why would the commanders at Lesbos even have considered the first alternative? From a weather standpoint, if Odysseus was correct in saying the winds carried him northward to Ismarus on Day Two, then the surface winds across the Aegean Sea would have been from the southeast—consequently, opposing (and thereby considerably lengthening) sea travel along the safer southerly route. The dilemma faced by the commanders would have been simply risking speed but danger (the northern route) against safety but delay (the southern route).

By Day Three out of Troy, the Greeks at Lesbos apparently decided on speed over safety, and began the northerly route directly across the Aegean Sea. Menelaus noted, "Then we asked Zeus to send an omen; he showed us a sign—commanded our ships to cleave the sea directly, head straight for Euboea—if we wished to flee most quickly from the threat of tragedy."[7] To experienced sailors, such a sign might be the beginnings of a strong north or northeasterly wind—the winds most favorable for a direct sea crossing. Since Greek gods often manifested their power through signs of nature, such an omen from Zeus would be reasonable. And, in fact, King Nestor noted: "A shill wind started up, the ships ran swiftly across the seaways rich with fish; that night we landed at Geraestus."[8]

Even today, that would be considered fast travel! Passage over the one hundred miles from Lesbos to Geraestus would translate to a speed of 7 knots for fourteen hours, or a more reasonable 5.5 knots for a longer travel day of eighteen hours. From a weather standpoint, the change from weak winds (Day One) to southeasterly (Day Two) to northerly or northeasterly (Day Three) might suggest the passage of a dry (no thunderstorms) cold front through the area.

Those three fleets continued to sail to Cape Sunium, near Athens—where Menelaus's helmsman died a sudden death. The fleets of Diomedes and Nestor continued on southward while Menelaus stayed to conduct funeral rites for his sailor. On Day Four, according to Homer, King Nestor commented that "the wind that favored me maintained its course; the god who sent it never sapped its force."[9] Weatherwise, that would imply that the northeasterly winds of Day Three continued into Day Four. So three of the fleets (those of Nestor, Menelaus, and Diomedes) reached the Greek mainland safely. What about the remaining three?

Agamemnon and his fellow captain, Ajax, having completed their appeasement rites (and not knowing the success of the first three fleets), likely chose the supposedly safer southerly route and headed toward Chios along the coast of Asia and so that they could begin island hopping. The sixth fleet, that of Odysseus's, likely had landed at Ismarus and begun a pillage of that ally of Troy—but was forced to flee by noon of Day Four. That fleet likely sailed westward along the coast of Thrace, perhaps stopping for funeral rites on the island of Thasos.

Day Five brought continued northeasterly winds so that the three fleets near the Greek mainland would have continued their voyages back to their respective kingdoms—Menelaus's fleet now far behind those of Diomedes and Nestor. By the end of Day Five, Diomedes likely had safely reached his homeland. Odysseus continued to travel westward along the southern coast of Thrace while Agamemnon and Ajax likely would have begun their island hopping through the eastern Cyclades.

Massive changes occurred on the sixth day from leaving Troy. First, it is likely that Nestor was able to complete his travels to Pythos on the west side of the Greek mainland. The others were not so fortunate. After the funeral rites for his helmsman, Menelaus likely would have resumed his travel southward to Cape Malea for his eventual return to his kingdom of Sparta. However, as the king reached the "high mountain of Maleiai," the most easterly of the three southerly promontories of the Peloponnesus, a powerful storm enveloped Menelaus's fleet.[10] As vividly described by Homer, Zeus "poured skill blasts and swollen combers, mountainous and vast"—the fleet was separated in the storm.[11]

Meanwhile, the fleets of Agamemnon and Ajax would likely have separated in the west Cyclades Islands by Day Six. Ajax's homeland of Lacadia lay to the north at the northwest tip of Euboea while Agamemnon's lay due west along the Arglikos Kolpos. The same storm that hit Menelaus's fleet now hit the ships of Agamemnon. Homer recorded that near Malea's promontory, Agamemnon experienced a "storm wind" that drove him "groaning heavily, off course."[12]

In a similar fashion, Ajax's fleet, now beating to the north, ran into the storm at an even greater angle than Agamemnon. Homer chronicled that "Poseidon first had dashed [Ajax's] ships against the giant rocks of Gyrae."[13] According to historians, the rocks of Gyrae have been variously identified as Mikonos (where the "grave of Ajax" is located), Tinos, or southeast of Euboea near the Capharean Promontory. All three locations are consistent with the occurrence of a specific weather phenomenon striking all of these ships. (We'll discuss that weather phenomenon in just a moment.)

The last of the Greek fleets that attempted to sail to Greece, that of Odysseus, also encountered a storm. According to Homer, the north wind, Boreas, "provoked by Zeus, who summons clouds, now swept against us: A ferocious tempest wrapped both land and sea."[14] Odysseus's ships were smashed by the storm; their sails were ripped and shredded. "In fear of death, we stowed our sails within our hold, then rowed and reached the coast."[15] That landfall likely would have been Euboea's northeast coast.

So what happened? It appears that on the sixth day out from Troy, four of the fleets encountered a massive storm—yet the four fleets were spread out widely across the Aegean Sea—Odysseus to the north, Ajax in the central Aegean, with Menelaus and Agamemnon to the south. How could one storm affect all four fleets?

Let's take a quick tour of one of the core theories of modern weather forecasting. During World War I, a group of Norwegian meteorologists proposed a brilliant theory to account for weather changes across Europe. That theory, called *polar front theory*, states

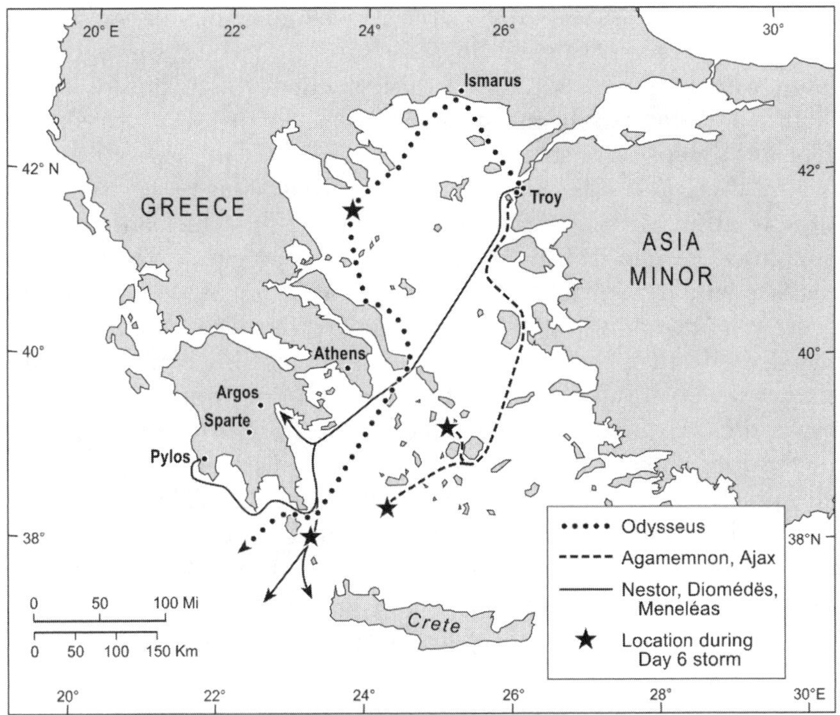

Map showing the locations of the six major fleets of the Greek navy following their departure from Troy. Map by Barbara Trapido-Lurie (adapted from Ref. 3).

that cyclonic storms form along the boundaries (fronts) between large areas having the same general humidity and temperature conditions (air masses). Storms evolve as the air mass advances into another. The part of the cyclonic storm that has cold air advancing into warm air is known as a *cold front* (because the Norwegian meteorologists were working in wartime and employed military terminology). A cold front is a long boundary—sometimes hundreds of miles long—that separates advancing cold air from another air mass of warm air. As the cold air advances into the warm air, the warm air rises, often creating a long line of thunderstorms. As we look at the map of the Aegean Sea with the locations of those four battered fleets, it is fairly easy to draw a long cold front extending from some supposed low-pressure center in the Macedonian mainland, south through the Aegean Sea over Odysseus's location; south through the region where Ajax wrecked and died; south through the location where Agamemnon's fleet was struck; and finally west to where Menelaus's ships were struck.

And therein lies the power of this analysis. This reconstruction of the weather associated with the voyages of the Greeks departing Troy shows a quite realistic meteorology. Not an event that modern scientists would consider "supernatural" or "miraculous" as having occurred in the immediate days following the Greeks' departure from Troy. Rather, the *Odyssey* shows a remarkably credible set of weather observations. So while this research does not conclusively answer the question of this portion of the *Odyssey*'s historical accuracy, it does show a consistency in descriptions of weather events and ship movements that is likely beyond coincidence. Since Homer and his fellow storytellers obviously knew nothing about polar front theory, this application of modern weather knowledge to ancient times provides an intriguing means of checking the possible "reality" of at least parts of the *Odyssey*.

THE MYSTERY OF THE ANCIENT GREEK DARK AGE

The Greeks under Agamemnon were ultimately victorious at Troy. Yet these great epics of Homer date to a much later time than 1300 BCE—perhaps as much as six hundred years later. Why was there such a lag between the events of the fall of Troy and their eventual recitation by Homer? We now know that something drastic happened to Mycenae culture a few decades after the fall of Troy. Most historians agree that by 1200 BCE, Mycenaean influence and power was clearly declining and within a few decades later absolutely collapsed. In fact, the archaeological ruins in Greece show that around 1250 BCE, all the glorious Homeric-era palaces of southern Greece were completely burned and destroyed, including the one at Mycenae. It was over five hundred years later that Greek civilization under the leadership of Athens and Sparta once again became a world power. After Mycenae, ancient Greece fell into a long decline, in essence, a dark age.

What happened to cause this rather definitive collapse of a great early civilization? In short, the answer is that we don't know for sure—but we do have some very interesting theories. One theory for the collapse that was favored by historians of the last century was that a people known as the Dorians invaded southern Greece from the north and brought down the Agamemnon culture. Now, however, that theory seems less acceptable for the very basic reason that there is no evidence of any invasion: The region appears to have been a relatively unpopulated area and nobody who spoke Dorian Greek moved into the area.

So the dark age of ancient Greece likely wasn't created by warfare.

Another theory suggests that the internal Mycenaean population turned on a weakened government and destroyed it, and then settled in regions formerly controlled by Mycenae. The theory goes that some of the displaced peoples escaped to Anatolia and elsewhere, where they gradually came to speak the Ionic dialect. But, if that were the case, why leave the relatively productive southern region of Greece? Strong archaeological evidence indicates that southern Greece from 1250 BCE to 750 BCE was depopulated. What could lead to such a massive depopulation of the area?

What if the region stopped being productive?

A theory first put forth by archaeologist Rhys Carpenter in the mid-1960s, and subsequently revived by three climatologists (Reid Bryson, H. H. Lamb, and D. L. Donley) in the 1970s, is that a massive long-lasting drought struck the area.[16] Such a drought, they reasoned, would have led to a growing frustration by the general population with the rulers of the region and promoted the unrest and eventual dispersal of the population.[17] Their evidence for such a drought? The climatologists used an interesting "analog" climate argument.

Let's take a closer look at that type of investigation.

Analog forecasting is a tried-and-true method used in weather forecasting. The fundamental premise of analog forecasting is that the current and future weather will behave the same way as it did in the past for similar conditions. The analog forecasting method says that if you are expecting a given situation—say, a strong cold front (like the one in polar front theory mentioned earlier) is forecast to move into the area—you would study all of the times when such a weather event happened in the past and see what the results of those situations were. Suppose—following our cold front example—you find that, for 70 percent of the situations in the past when a strong cold front passed through your area, severe thunderstorms developed. Using analog forecasting techniques, you would forecast that there is a good chance for severe thunderstorms with the current cold front passage.

The key to analog forecasting is to have enough past examples of a given weather situation that we can find a match to any given situation. Why do we need many examples? Fundamentally, it is virtually impossible to find an exact analogue to a given weather situation. Chaos plays a role. Every weather situation is unique to some degree and we have discovered that even small differences between the potential analogue and our forecast event can lead to very different results between the two. However, as time passes and the data of more weather events are stored, the chances of finding a good match improve. And, as we build a bigger database of events, we start to have greater confidence in finding the similarities—and differences—between potential analogues and the forecasted event. Also, with more events, we start to see the underlying weather pattern (say, a large high-pressure dome or a persistent low-pressure trough of air) that dominates the situation.

Climate forecast analogues work in much the same manner. Say, for example, we would like to make a forecast in spring for the upcoming summer's weather. To use climate analog forecasting, we might select the weather maps of spring seasons of the last hundred or so and group them into categories based on their rainfall, or maybe their temperatures. Once we did that, we would look at the average summer weather that followed each of those spring weather groups. We would take the weather map of our current spring season and see which of the groups were best matched to it. Then we would simply ascertain what was the normal summer weather that tended to follow that spring weather group. That weather, according to the analog method, would be the forecast for the upcoming summer.

The three climatologists, Bryson, Lamb, and Donley, followed a very similar procedure. First, using a seventeen-year database for rainfall and a measure of drought for seventeen stations in Greece, they conducted a statistical analysis to determine the common weather pattern of drought over the region. This was complicated by the fact that Greece commonly falls in a climate transition zone between drought and flood and, consequently, it demonstrates a bit more climatic variability than other regions of the Mediterranean.

Once the three climatologists derived specific weather patterns of drought and flood in modern Greece, they fitted the known movements of the ancient Greeks to a map of the country. Finally, they compared each of the known weather groups or patterns to that general migration map for the time of ancient Greece. Their reasoning was that if they could find similarities between the migrations and the occurrence of drought, it might help to explain why the ancient Greeks moved out of the area of Mycenae and to places like Anatolia. And they did discover an intriguing match.

Bryson, Lamb, and Donley found that the winter of 1954–55 produced a rainfall pattern across Europe that had, as a prominent feature, drought over southern Greece. And that particular drought closely corresponded to the pattern suggested by Carpenter as characteristic of the migration regions at the time of the Mycenaean decline. Now it is critical to note that our three illustrious climatologists stressed they had *not* "demonstrated that a drought of significant proportions did indeed occur in late Mycenaean times."[18] Rather, they stated that their analyses indicated the drought theory was the only hypothesis "presently available that is consistent with extant evidence."[19]

So the weather and climate of ancient Greece provide interesting clues in both the short-term and the long-term weather. First, based on the great Homeric epics of the *Iliad* and *Odyssey*, we may have indications of the actual day-to-day weather of those times over three thousand years ago. Nothing in the first part of Homer's *Odyssey*—the part describing the initial dispersal of the great Greek fleet after the sack of Troy—appears to invalidate known meteorological theory. Obviously, Homer didn't know anything about modern meteorological theory. The correspondence between the weather described in the *Odyssey* and modern weather theory provides an independent indication of authenticity to the epic story.

In a similar fashion, comparison of modern-day seasonal weather patterns with migration regions of the peoples in the times following the Mycenaean era of Greece provides intriguing clues as to some of the influencing origins of the five-hundred-year ancient Greek dark age.

Time: Approximately 1180 BCE
Location: The coast of the Aegean Sea, within a few miles of the ancient city of Mycenae, Greece, Europe

The boy had become a man, and now the man was nearing his ultimate fate. Hythamer's long, white hair fluttered in the onshore sea breeze that was strengthened by the annual Etesian winds from the northeast as he sat on a chair overlooking his estate on the Saronic Gulf. The old man's eyesight was starting to dim but he still stared out at the eastern sea with a yearning gaze. If only he could have been born in the heroic days of his long-dead grandfather. That great man—who had told Hythamer such fascinating stories when he was a youth—had been one of the renowned warriors who had risked life and limb to destroy the faraway kingdom of the horse-lovers so many long years ago. Yes, his illustrious grandfather had been one of those who had traveled across the great sea and lay siege to that notorious city of Troy. Ah, to have accompanied him! To have heard the sounds of battle, to have seen the gleaming walls of the legendary King Primus, to have fought under the command of the great Odysseus or Mycenae's own King Agamemnon . . .

Hythamer shook his head in sadness. Now such feats of heroism were simply the fodder of the increasingly fewer minstrels passing through the area for the religious festivals. No, instead, the times now were spent trying to simply survive in this sadly forsaken land. It was truly evident—as some of his fellow old-timers were saying—that the great gods had turned their favor away from Mycenae. Certainly the social situation demonstrated that hard fact—the fall of the great palace of Mycenae a few decades ago had shown that the great kingdom—despite the claims of that so-called king residing in a makeshift palace in the ruins of Mycenae—was long gone. No more did the people of the region look to Mycenae for protection and guidance. And to make matters worse, there was this continuing damn drought! Once again rains had failed to fall in sufficient quantity that year. It was likely, as Hythamer turned from the sea to look over the hillside rows of scraggily brown grape plants, that the wine harvest would be bad again this year—perhaps the worst yet.

Hythamer heaved a tired sigh. The brilliant blue cloudless sky foretold that today—as had so many days of his life—would be once again be dry and warm. If only he had been born in the heroic days of his grandfather's king . . . the great Agamemnon.

Chapter 8

THE MYSTERY OF PETRA, THE ROSE CITY

As we move into the days of the Roman Empire, we find many intriguing similarities between the climate of Julius Caesar and that of today. One pleasing aspect of climate study in Roman times is that our historical record becomes much more detailed than for earlier times, and we can start to examine the weather of individual seasons or, in some cases, even specific storms. And that research tells us that, all in all, the individual weather of critical historical events in Roman days would not be considered abnormal in today's world.

For example, weather figured prominently in Caesar's legendary second campaign into the British Isles in 54 BCE. Throughout the long summer, the Romans were forced to wait for many weeks while west-to-northwest winds persistently blew across the English Channel, rendering sailing impossible. Today, we know that such constant wind would be the result of a very stable weather situation; in this case, a very large high-pressure system extending from Great

Britain south to the Azores. And this high-pressure system would have other effects—it would tend to force storms northward, keeping them from reaching Western Europe. A weather detective would deduce that perhaps then Western Europe would be dry. As verification of this climate inference, Caesar wrote in *De Bello Gallico* that his supply problems were aggravated by a very small harvest of autumn crops on account of a marked dryness.[1]

Finally the winds of that dry summer in 54 BCE turned. Today we know that would mean the blocking high-pressure system had broken down—and its demise would allow storms back into the area. Consequently, should we be surprised about what happened after the Romans landed and established a beachhead near Dover? Immediately after they had pulled their war galleys onto the beach with the transports anchored offshore, a destructive evening thunderstorm struck. The severe storm created high waves that wrecked the beached galleys and even drove the anchored transports onto the rocky shore. All in all, twelve Roman ships were destroyed and, because of this loss, Caesar was forced to retreat back to France.

Specific weather can definitely play a determining role in history.

However, the question remains as to the impact of long-term weather—climate—on the aspects of Roman civilization. Were Rome and its client states across the Mediterranean influenced by the climate of the time?

As an answer, let's look at one of the intriguing climate mysteries of the Roman age. It involves a beautiful abandoned desert city known as Petra located in the stark and arid Jordanian desert. If you have seen the movie *Indiana Jones and the Last Crusade*, you have glimpsed a part of Petra—specifically, the magnificent rock-hewn "Cathedral of the Holy Grail" (actually a building commonly called the Treasury). Shown at the end of the movie, Indy and his companions escape almost certain death from the Treasury before riding off in triumph down a narrow canyon.

But Petra was much more than the closing scenery of a famous movie.

Time: 31 CE
Location: The city of Rekem (known in modern times as Petra) in the central Jordanian desert, Asia

Betad stopped along the paved street of the city between its magnificent colonnades and turned to his guest, a distinguished visitor from Rome.

"I am glad that you have had the opportunity to behold the sights of Rekem, Senator. Few visitors ever forget it," Betad said with a certain pride.

Senator Rufus Gaius Metallus, visibly awestruck, turned in a slow circle, trying unsuccessfully to experience the panorama in a single view. He had arrived on camelback from Gaza at sunset only yesterday and so the shadows of the setting sun had hidden the glorious natural and human-made splendors of the city.

This morning, as the red-haired Roman turned his gaze around the city, signs of incredible wealth paraded about him. He spied literally hundreds of workmen laboriously carving several majestic rock-hewn buildings that protruded out of the sheer canyon walls. From atop rock walls separating large estate residences, the lush and green vegetation poked out: visible indications of tens of small hidden gardens. Even his ears could detect the wealth of this city; the faint sound of water trickling through channels echoed softly somewhere behind the walls.

"I can't believe the massive amount of riches here. Those stone works alone," Metallus said, pointing a manicured finger west toward the rock-face mammoth carving into the sandstone cliffs, "are worthy of those found along the Sacred Via in Rome itself."

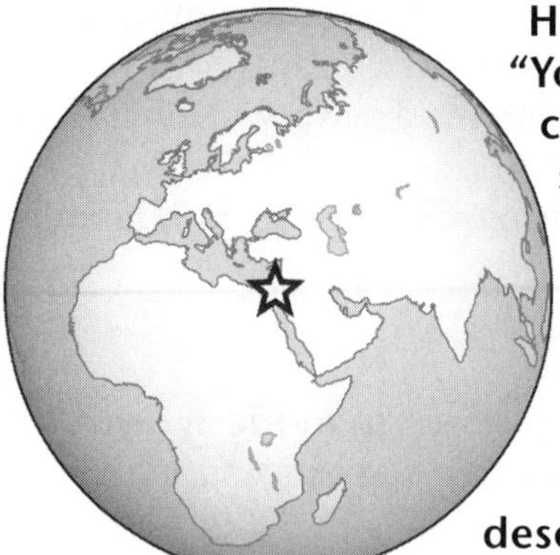

He shook his head in admiration. "You Nabataeans appear to have created a paradise out of utter sunburnt desolation!"

In response to the growing morning heat, Metallus took a small cloth and wiped off his sweaty forehead.

Betad nodded.

"Indeed, Senator, we of the desert have been blessed by the

gods," the Nabataean merchant replied in humble fashion. Silently, he added to himself, "And due in large part to the hard gold and silver coinage we charge your Imperial caravans, praise to Atargatis, dolphin goddess of trade."

"So who lives in those rock palaces?" the senator asked, pointing up toward the ornately decorated buildings carved out of the rock. "Your king and the royal family?"

Betad shook his head. "No, they aren't palaces—well, not for us, anyway—although the royal quarters are quite impressive and I will be taking you there in a moment to the court of King Aretas, the fourth of our kings to bear that name. No, those rock temples are for our gods and our illustrious ancestors—in particular, they are monuments to our forebearers."

Rufus Gaius Metallus raised his thick red eyebrows without comment. The dead of Rome were, with few exceptions, cremated. To memorialize the dead seemed rather wasteful to the Roman patrician; nothing compared to the commemoration of a military victory.

"So," Metallus finally said, shifting the topic back to business. "How much will you charge to feed and water my caravan of spices coming in from Damascus? I have just received word that it will arrive within a few days. Obviously we will need to quickly refresh the camels and porters here before continuing on to the Mediterranean through Gaza. My chief servant estimates that they will need to rest here for at least three days."

Betad smiled, pointing to a long line of camels that was departing east into the desert, past the huge amphitheater under construction. "With the Greek merchant Orsetes's caravan leaving today, my stables will have more than enough room to handle your needs. As to the costs . . ." He coughed in apparent embarrassment. "I'm afraid the costs demanded by our royals have increased slightly since your last caravan."

"Indeed," the Roman said dryly. "The cost of

building your enormous ancestral temples is proving a bit expensive?"

Betad smiled. "Indeed we humble merchants here in Rekem must struggle to earn a poor living from this inhospitable desert."

Petra wasn't the name that the inhabitants of that great lost city gave it throughout its long rise and eventual fall.[2] Some ancient writings, such as the Dead Sea Scrolls, cite the name Rekem (or Reqem) as Petra's actual name when it was in its prime. Others suggest that the city name was Sela or Sella, after the Edomite site atop the nearby mountain of Umm al Biyara. What is true—no matter what its original name—is that Petra (that later moniker derived from the Greek word for rock) is a magnificent archaeological site in the southwest deserts of Jordan. It is renowned for the ornate stone buildings carved directly from its surrounding rock canyons. The noted Hebrew scholar Josephus and the Greek geographer Strabo, for example, both mentioned the famous "rose rock" desert city. The emperor Hadrian actually visited Petra and declared it a provincial capital. Yet this opulent city was abandoned—all but vanished to Westerners—in the final years of the Roman Empire. Great Petra remained lost to the world until its relatively recent rediscovery by the Swiss explorer Johann Burckhardt in 1812. Burckhardt uncovered the once-great city while traveling between Damascus and Cairo. He was on a search to discover the lost city in order to ascertain the validity of legends that placed Petra at the foot of Jebel Haroun, the mountain burial site of Aaron, the brother of Moses, leader of the Exodus.

In Roman times, however, Petra's fame—and its incredible wealth—were not linked to the Bible but rather to its economically strategic, geographic position. It was centrally located at the crossroads of a number of caravan routes from Gaza to the west, Damascus to the north, Egypt and the Persian Gulf to the west and south, respectively. Occupied by a formerly nomadic people called the Nabataeans who had moved into the region, Petra was a well-established city by the third century BCE, and had become a major center of commerce by the first century BCE.[3] The Nabataeans were an Aramaic-speaking Semite people who moved into the area sometime after the Babylonian destruction of Jerusalem. One of the earliest historical references to the Nabataeans at Petra is with regard to a successful battle they undertook against one of Alexander the Great's generals in 312 BCE.

By virtue of their strategic location at the crossroads of the lucrative silk trade route from the Far East and a profitable transport of spices and fragrances from southern Arabia, the Nabataeans flourished and prospered. By the time of King Aretas IV, the Nabataean kingdom was bounded by Gaza to the west, the Arabian Gulf to the south, and Syria on the east to Hagra on the west. Following Roman expansion into the area, the rich kingdom morphed into a puppet state of the empire. With the death of King Rabel II in 106 CE, the Nabataean kingdom was integrated into the Roman *Provincia Arabia*, and, as a tacit acknowledgment of its wealth and power in the region, Petra became the provincial capital. That dominance was further exemplified in 131 CE when the Roman emperor Hadrian visited the city.

The critical question from the standpoint of climate is how could a civilization survive—and even flourish—in what is today a very sparsely populated arid desert? In Roman as well as modern times, the countryside of southwestern Jordan near Petra was a beautiful but stark landscape of red sandstone. Erosion and weathering had produced steep-walled wadis (gorges or arroyos) and towering sandstone pillars and buttes. The city of Petra itself developed in a roughly crescent-shaped valley, surrounded by steep cliffs rising to three hundred feet or more above the valley floor. The tops of the surrounding plateaus had only limited plant life—most of the vegetation was restricted to the valley floors.

Petra's present climate is harsh and unforgiving. The region even today averages at most between five and six inches of rain a year. Most of that rain occurs in the winter. The summers are generally characterized by markedly hot and dry conditions. The beginning and ending of the hot summers are normally heralded by arid, sometimes gale-force winds from the east or southeast. These seasonal winds, regionally known as the *hamsin* (sometimes written as *khamsin*, or *khamseen*), can pick up huge amounts of dust and actually lead to even hotter temperatures over the area.

Conversely, the winters can produce—in the context of the region's characteristic aridity—torrentially strong frontal storms that pound the desert rock and pavement. These rains normally fall from only a relatively few number of storms per year. Consequently, flash flooding from these storms is common for the area. Walls of some of the canyons of this region are carved (and reshaped with each year's storms) by these intense, but short-lived rains.

What is particularly interesting is that we don't think these modern climate conditions are much different than those of ancient Petra. So how could the Nabataean civilization not only survive under those harsh, dry desert conditions, but actually thrive? The answer doesn't lie in the climate but rather in three human artifacts: money, engineering, and an ethic of sustainability.

Those ancient Nabataeans had those three highly significant factors in their favor in their quest to create and maintain a thriving kingdom in the midst of a formidable desert. First, they had almost unlimited resources of money, garnered from their caravan crossroads trade, to use to improve their lives. Second, and equally important, they had some very talented engineers with the genius to overcome a variety of very difficult technical water problems. And, third and perhaps most crucial, they had fostered a strong desire to overcome the natural limitations of their environment.

If the desert didn't have a nice even distribution of rain spread throughout the year, then, the Nabataeans through money and genius would simply create the means to *hold* that water. Throughout this arid desert of southwest Jordan, the ancient Nabataeans proceeded to create a complex—and costly—masterpiece of hydraulic engineering.[4] Most of the work involved waters from the bountiful Ain Musa (Spring of Moses). Over the course of several hundred years, they constructed and maintained an immense water conservation system that included the creation of a series of floodwater retention and diversion dams, each designed to divert and hold the rush of swollen winter waters to prevent flash floods.

This was accomplished by channeling and diverting water from the normally dry wadi, sandy streambeds, in which the city was constructed. These ancient engineers also lined the adjoining so-called Siq with water channels. The Siq is a deep gorge or extremely large chasm that leads from the outside desert to the entrance of the old city (and the narrow canyon that Indiana Jones and his friends rode through in the movie). This system of channels and dams controlled water flow and erosion, and provided irrigation water to the settlement. Indeed, satellite and photographic imagery have also revealed the Nabataeans' incredible genius in hydraulic engineering in the manner in which the wonderful rock-sculpted buildings were cleverly carved to facilitate the natural flow of rainwater. The Nabataeans created such an impressive system of overcoming the environmental constraints of the desert that, at its height, archaeologists estimate that Petra may have held a population of twenty thousand in comparative luxury in the desert.

Indeed, by the time the Romans seized direct control of the desert metropolis at the start of the second century CE, the Nabataeans had in place an intricate water system that included hand-carved stone flumes (some even lined with ceramic pipes), reservoirs, and over two hundred water-holding cisterns. This system is estimated by scientists today to have been capable of supplying as much as twelve million gallons of water a day to the residents of the prosperous desert urban center. Archaeologists estimate that the water system was so colossal that it could have met the needs of a hundred thousand people in a modern-day American city. Even with a smaller population, the "engineered" water of Petra gave life to numerous gardens and animals spread throughout the city. In essence, that carefully tended water was the foundation of a rich urban culture in the middle of an inhospitable desert.

So what happened? Today, Petra is an abandoned desert ruin—no gardens, no water . . . and no people. What caused that wonderfully engineered city to collapse?

In this case, "collapse" is literally the word to use. The geology of Petra and its surroundings are dominated by faulting—breaks that delineate the movement of one rock surface against another. On May 19, 363 CE, those rock surfaces abruptly released their stored energy and unexpectedly moved.[5] That crustal earth movement created one of Jordan's most destructive earthquakes, damaging and destroying hundreds of cities in the region, including Petra.

In Petra, the massive stone columns that had lined the city's main thoroughfare were toppled by the earthquake and its aftershocks. And, most devastatingly, that earthquake severely damaged the magnificent hydraulic infrastructure so carefully constructed by the Nabataeans over the previous centuries. That, coupled with shifts in the desert trade routes over the previous century, left the community without the resources to rebuild or maintain what structures were left. Without any maintenance, Petra's water system slowly crumbled. The Nabataean dams and canals no longer diverted water flow away from the ancient monument and the ruins of the town. Flash floods began to wear away at the very rocks of the decaying city. Scientists, such as geomorphologist Dr. Tom Paradise of the University of Arkansas, have carefully documented the slow weathering of the historic buildings of Petra over those long centuries.[6]

Petra became a ghost town, and then faded into the mists of a legend. After five hundred years of being a bastion, showcasing the power of humanity against the desert's hostile environment, the city reverted back to the "natural" conditions of the wilderness.

Are there any fundamental conclusions that we might extract from Petra's engineered climate mystery?

Interestingly, a modern analogue to the Nabataeans may exist in the United States today. I live in one of the suburbs of Phoenix, Arizona, a desert megalopolis of over four million people. This city exists in an arid environment without any huge natural lake or consistently flowing river located throughout the entire metropolitan area. The climate of the central Arizona Sonoran Desert produces roughly six to eight inches of rain a year. In essence, in terms of its climate and environmental setting, Phoenix could easily be called an "American Petra." But it also earns that appellation because it had those three key attributes that allowed Petra to exist and flourish for centuries in the arid central Jordanian desert: money, smart people, and an inherent desire to overcome the natural limitations of the environment.

The engineering marvel created by those three attributes is today called the Salt River Project, or SRP. At the beginning of the last century, Phoenix was a very small agriculture-based town of only a few thousand people. Those residents with incredible foresight established SRP as a safeguard against the harsh desert environment. Water in the Desert Southwest—particularly for irrigating crops—can be very scarce. Indeed, at the time of the creation of SRP, the area was actually undergoing a very bad drought, even for a natural desert area. The water supply from natural flowing rivers had become unreliable. Consequently, in 1903, the area's farmers and business leaders formed the Salt River Valley Water Users' Association. They pledged more than two hundred thousand acres of their land as collateral against a government loan to build—and maintain—a huge water storage and delivery system.

That loan provided, for the time it was awarded, massive funding for the construction of water storage dams and canals, which, eventually, led to subsequent major development of the area. Interestingly, the modern engineers often utilized the very same canals first dug by another lost civilization, the Hohokam. One of the first of the eastern water-retention/diversion dams—a mammoth structure named after President Theodore Roosevelt—was built on the Salt River, a stream that flowed somewhat intermediately throughout the year and whose water was based on snowmelt in the high mountains to the east in Arizona. Since that time, engineers at SRP have added three more dams on the Salt River and two dams on the Verde River to complete a thirteen-hundred-mile system of canals and laterals. As more and more people have migrated to Phoenix, the system that first was developed for agricultural purposes has become the central means for supplying water for the mushrooming urban population.

Aerial image of the Roosevelt Dam, a water-retention reservoir for the Phoenix metropolitan area, on the Salt River in central Arizona. Photograph courtesy of the Salt River Project.

Just like Petra, the creation of a reliable water storage and delivery system brought new life to the desert land. Experts agree that, more than any other single factor, the SRP water system has influenced the Desert Southwest's development.

What often intrigues me is the argumentative oddity of some of the people who stand in wonder of Petra's justifiably magnificent water system. Some people who applaud the genius of the Nabataeans in overcoming their environment and creating such a glorious city are often the same people who loudly denounce the creation and maintenance of water management systems such as those of SRP and those on the Colorado River, such as Glen Canyon Dam. They protest that such hydraulic systems are complete "abominations of nature" and should be torn down. I am reminded of a saying from a favorite author of mine, Robert A. Heinlein: "There are hidden contradictions in the minds of people who 'love Nature' while deploring the 'artificialities' with which 'Man has spoiled 'Nature.' The obvious contradiction lies in their choice of words, which imply that Man and his artifacts are not part of 'Nature'—but beavers and their dams are."[7]

Perhaps Petra teaches multiple lessons. First, both Petra and the modern society in the Southwest United States have utilized two of our greatest human resources—our intelligence and our wealth—to allow for a better lifestyle than previously existed in those places. As a resident in the quite-artificially maintained environment of Phoenix, Arizona, I am rather selfishly pleased that those resources have been available for my enjoyment and lifestyle in the Southwest United States.

But it is now the third attribute of Petra's environmental philosophy, the inherent desire to overcome the natural limitations of the environment by "sustainability," which is currently being debated in Western society. Do we have the moral right to modify and extend the environment to our own needs? Many today answer with a resounding no and say case closed.

But, if that does indeed become the will of the majority, do we then have the will to cope with the consequences of that moral judgment? If, for example, power, water, and flood protection are not available to a given population in need of it because the larger society wants to protect the environment, can the larger society afford to sustain that given population somewhere else or in another manner?

Every action—and every moral judgment—has consequences and costs.

Petra's second lesson is that even the best climate/environmental engineering systems can fail over a long enough time. Petra flourished for over five hundred years—more than twice the current age of the United States. Societies do change and the ability of those changing societies to adjust their environments is perhaps one way to gauge the success of a society. It seems to me that, when we study history, we often are perhaps overly concerned with the eventual failure of a civilization and neglect to appreciate the long extent to which it survived. With Petra, the Nabataeans met the challenge of maintaining a unique, prosperous lifestyle in a naturally hostile environment for a very prolonged period of time. But eventually they succumbed to the vulnerabilities of their engineering. Is that success or failure?

Time: 31 CE
Location: The city of Rekem (known in modern times as Petra) in the central Jordan desert, Asia

Betad watched the Roman senator's long camel caravan slowly depart his walled hostel, as he held a quite large bag of gold coins in his hands. Glancing down at the bag, Betad quietly intoned a short prayer to the goddess Atargatis. Undoubtedly, the senator would recoup—and likely greatly exceed—those costs by getting his Arabian spices and perfumes back to Rome. The senator had explained that most of the rich patricians in Rome headed out of the city during the heat of summer for the more temperate seaside resorts such as Herculaneum. If his caravan reached Gaza and then sailed back to Rome by summer's end, the senator's clients would be able to sell their delights to the returning sweating masses quickly.

Betad shrugged. It made no difference to him. Despite how the senator had continually harped on the "glories of Rome," Betad knew that he lived in truly one of the best places and times on Earth. He glanced at the rising sun's glow on the buildings that lined the Siq and smiled. Yes, indeed, this was a beautiful city.

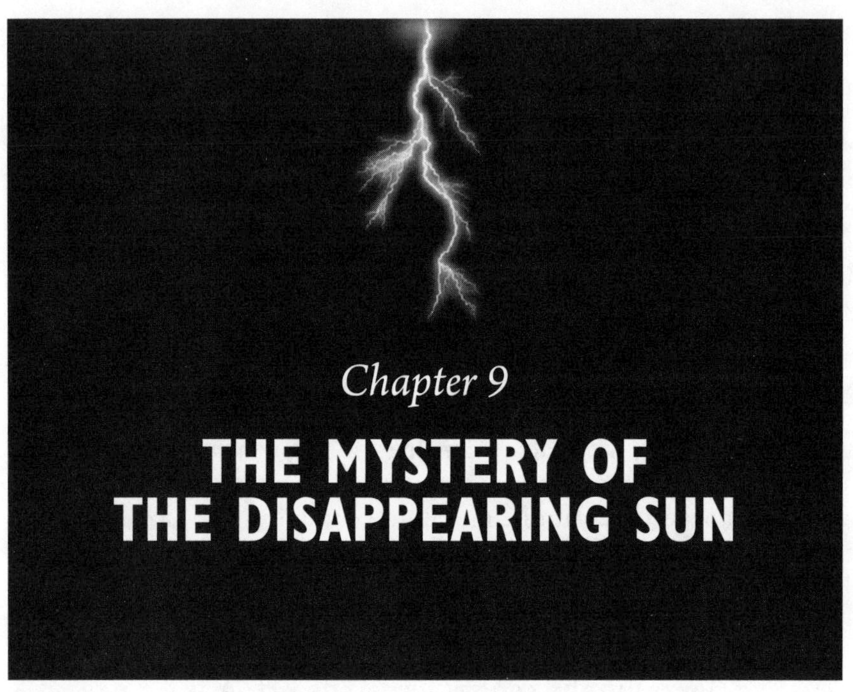

Chapter 9

THE MYSTERY OF THE DISAPPEARING SUN

Something very odd happened around the year 536 CE. That much is known for certain. History tells us that at the very least many people particularly around the Mediterranean Sea died in a very gruesome fashion. Rather than try my hand at writing a fictional perspective of the years immediately after 536 CE, I'll let a historian by the name of Reverend Thomas Short describe vividly the problems as written in his 1749 book, *A General Chronological History of the Air, Weather, Seasons, Meteors, etc. in Sundary Places and Different Times.*[1] Reverend Short is paraphrasing the Byzantine historian Procopius, who lived in Rome at the time.

Time: 536 CE
Location: Rome, Italy, Europe

On the 14th Day before Calends of March, the Sun was eclipsed from Morning till three a-Clock in the After-

noon—The Lands of Italy laid uncultivated last Year, hence a great Famine. Such as dwelt in Emilia left their Seats and Goods, and went into Picenum, and even there no less than 50,000 died of Famine. Then the Starved throwing off all Humanity, killed and eat one another. Delicate Mothers eat their tender Babes. Two Women killed 17 Men and eat them. A woman in Milan eat her dead Son. People kneeling down on their Knees and Hands to eat Grass and Herbs, fell down with Weakness and died, nor was there any to bury them. Others eat Dogs, Mice, Cats and the vilest Animals.

The Disease spread as among great Herds of Cattle. Their Bile was redundant, there was no Juice left their Bodies. Their Skin was hardened, and became dried like Leather, and clave to the Bones, their livid Colour became black, Men looked like Charcoal Wood, their Countenance was senseless and stern. They died every where, partly from Hunger and partly from too great Satiety. Having been burnt up within, after the natural Heat was extinguished. For having been starved, if they had any Opportunity to feed freely, being not able to digest their Food, they died so much sooner.

What could have caused such a terrible famine and plague? Estimates place the dead for the first few years after 536 CE at perhaps three hundred thousand people. Many sources indicate that the plague—eventually called the Justinian plague, one of the first Western occurrences of the infamous Black Death, which started in Egypt—then made its way around the Mediterranean Sea before eventually ravaging Europe. And then it got much, much worse. That plague led to several others that inflicted Europe over the next several decades from 542 to 565 CE. By its end millions of people had died—some say as many as a third of Europe's total population at the time succumbed to the plague and its related problems.

Did climate change do that—and do it that quickly?

As a climatologist, I'll give a short definitive answer: No, not in this case.

What? Isn't this a book about climate and weather change and their effect on civilization?

Fundamentally, disease isn't caused by climate and weather—but they can definitely play a role in setting up the right environment for disease, particularly its rate of dispersion. If the proper climatic conditions exist, disease can be spread rapidly. Noted climatologist H. H. Lamb observed that "the incidence of the disease in the Middle Ages seems to have been worst where it was concentrated in locally warm and moist habitats in cities and in other concentrations of populations and contacts along the routes of travel."[2] Climate and climate change cannot *cause* disease but they can definitely influence its development. But before we look at the climate of 536 CE and after, let's investigate some of the other oddities that abounded at that strange time in history.

Volcanologist R. B. Stothers examined the historical records of 536 CE and found four interesting firsthand accounts. First, as with Reverend Short above, Dr. Stothers also quoted Procopius for the year 536 CE: "The Sun gave forth its light without brightness, like the Moon, during this whole year, and it seemed exceedingly like the Sun in eclipse, for the beams it shed were not clear nor such as it is accustomed to shed."[3] Another chronologist living in Constantinople in Turkey wrote, "the Sun became dim in the course of the recently passed fourteenth indiction, for nearly a whole year, when Belisarius [a military hero] had the highest honour, so that the fruits were killed at an unseasonable time."[4] Another inhabitant of Constantinople wrote, "The Sun began to be darkened by day and the Moon by night . . . from the 24th of March in this year till the 24th of June in the following year."[5] Still another resident in the region wrote, "The Sun was dark and its darkness lasted for eighteen months; each day it shone for about four hours, and still this light was only a feeble shadow . . . the fruits did not ripen and the wine tasted like sour grapes."[6]

These four sources all suggest that something happened to greatly obscure the sun for perhaps as long as fourteen months. The incident has been termed a *dry fog*—a haze of suspended dust rather than suspended water droplets. Dry fogs tend to develop quickly and last for a long time. For example, a fog of this type occurred in 1783 and extended across most of the Northern Hemisphere. Dry fogs, as noted by the natural scientist T. L. Phipson, "exhaled a disagreeable odour,"[7] and were remarkably dry compared to normal fogs.

But what causes a dry fog? A nineteenth-century meteorologist named L. F. Kaemtz suggested that the dry fogs in 1834 resulted from the burning of "heaps of bad grass and potato-culms [plants]."[8] Others suggested that Earth periodically passes through the tails of comets and the comet dust created the dry fogs. Interestingly, the first person to get it right was none other than the great American scholar Benjamin Franklin. Franklin theorized—correctly—that the 1783 dry fog was the result of a large eruption of an Icelandic volcano. He suggested that the remote volcano pumped huge quantities of dust high up into the atmosphere, and it was that volcanic dust suspended in the air that accounted for the global haze.

Today most scientists tend to agree with Franklin: most dry fogs occur when large volcanic eruptions eject massive quantities of sulfuric gases and dust into the high atmosphere. If those materials get into the stratosphere, the region where the ozone layer exists, an atmospheric haze can last for years—and consequently affect climate and weather.

Research now shows that large, sulfur-laden eruptions of volcanoes can lead to widespread cooling for one to three years after the eruption. However, these are the cases where the ejecta—the ash and dust—are pushed high up into the levels of the atmosphere beyond where jet planes fly. Mt. St. Helens, with its more horizontal-oriented eruption in 1980, did not produce significant global cooling. However, Mt. Pinatubo in 1991 did have a strong detectable effect on global temperatures for several years after its eruption.

Was that the case in 536 CE?

Our best indicators of climate at that time are three distinct sources: contemporary historical accounts such as those by Procopius and two physical measures of climate—tree ring analysis (dendrochronology) and ice core analysis. Both of these have the great advantages of being very precise in their ability to date to specific years and also of being quite reliable in mirroring the temperature and precipitation changes. For the mystery of 536 CE, let's take an in-depth look at one of those investigative tools—tree ring analysis, or dendrochronology. Interestingly, the science of using trees to explore past climate is an incredibly young field of study—and, in contrast to many sciences, its founder was a single individual.

A. E. Douglass is the true "father of tree ring research," the scientist credited with creating the field of dendrochronology. Indeed, he even coined the name dendrochronology, literally "tree timekeeping," for the discipline. Douglass began his career as an astronomer working in the high plateau country of northern Arizona. At the turn of the twentieth century, the young astronomer identified a possible relationship between climate and the growth of certain trees. So, almost as a hobby, he began documenting the annual rings of pine trees and Douglas firs in the area.

One of the first things that he noted was the timescale of tree rings is "absolute"; that is, individual years can be identified and, in fact, in some cases, individual seasons within a year can be precisely seen in the tree rings. As he discovered more characteristics about tree rings, he realized that a tree didn't need to be cut down in order to document and study its rings. Today's modern tree corers do not

Tree coring of a pine tree by the author using an increment bore in northern Arizona.

harm the living tree. Instead, pencil-wide cylinders of wood are extracted using increment bores from two or more sides of ten to twenty trees, all of which are growing on the same kind of site.[9] The trees can quickly recover from such a small extraction of wood.

After coring many samples and studying them in the laboratory, A. E. Douglass realized that the width and character of the rings of many trees in the area were demonstrating the same characteristics of growth from year to year. He noted that widely spaced rings for trees of many species, including the mammoth Sequoias in California, were generally produced during wet years, while narrow-spaced rings appeared to represent dry years.

Since that time, we have refined exactly which species—and, equally important, which environmental conditions—create the best tree ring records of climate. In terms of species, long-lived trees tend to display the best well-defined growth rings. For example, in the western United States, these tree species include the Douglas fir, bigcone spruce, ponderosa pine, pinyon pine, limber pine, bristlecone pine, Jeffrey pine, and white firs, to name a few. In fact, in the White Mountains of California, a bristlecone pine tree, nicknamed "Methuselah," has been determined to be over 4,700 years old—it was a seedling at the time of the building of the Egyptian pyramids![10]

The other critical factor for the climate analysis of trees is the specific environment that surrounds a given tree. We have found that the tree displays the best climatic variations in its ring growth if it periodically experiences growth stress from climate. For example, a tree in the middle of its natural habitat will probably not display the best record of climate variations, while a tree that is at the edge of its habitat—and, consequently, undergoes greater stresses such as droughts and freezes than its interior counterparts—will have a growth record that better shows climate variations. This idea is called the *limiting conditions theorem* by dendrochronologists and implies that site selection is one of the critical aspects of tree ring science. Sites have to be carefully selected to maximize the amount of climatic variability seen in the tree's ring widths.

Trees growing on sites where climate seldom limits growth, such as in the center of large forests, produce rings that are generally uniformly wide. Such trees are called complacent and do not make good climate indicators. In contrast, trees growing on sites where climatic factors are more variable, such as on slopes or on rocky terrain, produce rings that can vary greatly in width from year to year depending on how severely the climate has affected growth. These trees are called sensitive and are exactly the type of trees that we want to sample.

So Douglass determined that the record of growth seen in tree rings could be used as a "proxy," or substitute, for actual temperature records. That was exciting—but worked only for areas where there were extremely long-lived trees. Was there any way to extend the climate record created by trees far back into time?

The eureka moment for Douglass came in 1909 when he joined an archeological survey of the Southwest United States. He noted that many of the timbers used in the ancient pueblo ruins throughout the region showed the same kind of growth-ring variations that he had seen in the living trees he had sampled earlier. The all-important critical link between the two sets of tree rings occurred in 1918 as he was examining a roof beam taken from a pueblo near the town of Show Low in the eastern part of Arizona. He discovered that growth rings from that piece of wood bridged the time gap between the older rings of the other Arizona pueblos and the newer rings of living trees—thus he suddenly had a complete record of climate back to 700 CE! He discovered one of the fundamental concepts in dendrochronology: cross-dating.[11]

Cross-dating is a complex challenge in puzzle-matching. What a researcher must do is to find the rings at the early end of one tree core that correspond to the rings at the late end of another—and match that one core's early rings to still another's later rings. With enough samples, the record created from these overlapping tree rings can be extended back centuries. So from Douglass to today's researchers, literally thousands of trees have been cored across the world—and we have begun to piece together, through cross-dating, a long time line of climate changes for many places. In particular, for the Mystery of the Disappearing Sun, we begin to see exactly what was happening around the world after 536 CE.

The tree ring scientist who was most intrigued with this odd "solar dimming" was Mike Baillie, a professor at the Queen's University of Belfast in Northern Ireland.[12] After compiling most of the tree ring records for that time period, he made a rather startling conclusion: almost every tree ring analysis across northern Europe, Asia, North America, and even South America showed markedly reduced growth for the years immediately following 536 CE, implying significantly colder conditions. In particular, the majority of European tree ring analyses from Ireland to Sweden to Germany showed one consistent signal for the years just after 536 CE: the trees were having a hard time. They simply weren't growing much. For example, a composite of all oak tree growth across Europe indicated that growth dramatically dropped to 85 percent of normal growth in 536 CE, followed by even more dramatic drops to 75 percent of normal growth by 540 CE. Indeed, most of the world's tree ring records suggest that 540 CE was one of the coldest years to have occurred in the last two thousand years!

Okay, the trees say it was cold. And, as mentioned above, historical written records such as those by Procopius confirm those physical findings—it was dark and cold. With two confirming sources, some might be convinced. But many scientists have an obsessive-compulsive nature—they like slam-dunk situations. While the tree ring record and the historical record both show that it was cold, is there any other evidence to confirm that finding? After all, three indicators are even better than two. In this case, the researchers turned to ice core analysis.

As discussed earlier, a powerful tool for investigating climate change is ice core analysis. We know that ice forms in nice annual layers on ice caps like Greenland—one year's snows falling on the slowly ice-morphing snows of previous years, creating ice layers that date back thousands of years. Consequently, if we drill into the ice, we can extract cores—in many ways similar to tree ring cores—that tell us the climate with precision over those thousands of years. So, if we study the ice core record from Greenland, will it also suggest the occurrence of an odd event around 536 CE?

The original analysis of the best Greenland ice core prior to the 1990s identified a very high sulfuric acid signal in sections of the Greenland ice cores dating to 536 CE. As I mentioned with the supervolcano Toba, high levels of sulfuric acid have been associated with marked volcanic cooling. That was promising—all the evidence seemed to be pointing to a major volcanic eruption that put enough ash into the air to dramatically cool the climate for a few years—just like the Pinatubo eruption did to world climate a few years ago. In fact, a leading scientist wrote an article in 1984 for the prestigious science journal *Nature* in which he stated exactly that point: the 536

CE dry fog event was the result of a large volcanic eruption. He did say, however, that "the [actual] volcano responsible remains a mystery."[13] So can we say the climate mystery is solved, except for a tiny bit of detective work about identifying the actual volcano?

We have a potential problem. In the mid-1980s, glacial research scientists conducted an exhaustive reanalysis of the Greenland ice core data—and when they finished, they said that the volcanic dust they found in the ice core dated back to 506 CE, not 536 CE. And that earlier date would be too early to cause the serious climate problems linked to the Justinian plague. Subsequent studies of the ice cores from Greenland reconfirmed the analysis. So the ice core scientists concluded rather definitively by the mid-1990s that there was no major volcanic eruption in 536 CE.[14]

But remember our tree ring analysis and written history both agree that something big and nasty did indeed happen around 536 CE. If it apparently wasn't a volcanic eruption, what was it?

In the mid-1980s, tree ring researcher Mike Baillie proposed a rather cataclysmic idea.[15] He suggested an asteroid or comet—much smaller than the "dinosaur-killer" comet discussed earlier—might have impacted the earth. At the time most scientists dismissed his theory. Scientists have generally disliked employing cataclysmic theories as explanations because such theories are simply very hard to prove. If something has happened only once or twice, it is much harder to see consistent signs in the climate record than if it happened over and over again.

Scientists changed their minds in 1994. In that year, a large comet—named Shoemaker-Levy 9, after its discoverers—broke up and impacted the planet Jupiter, quite literally giving the kingly planet a pounding. The Hubble Space Telescope documented a series of huge black splotches that were created in the Jovian atmosphere by the plummeting bits of space debris. Scientists abruptly realized that cataclysmic events could indeed occur in our solar system. As they studied the event, astronomers computed the average size of the multiple fragments of the comet as they impacted Jupiter. They were on the order of about three hundred yards to perhaps a mile or so in diameter. So they were much smaller than the "dinosaur-killer" asteroid of sixty-five million years ago.

That determination led to an interesting question: What if a smaller comet—not a "planet-buster" but perhaps something smaller than a Shoemaker-Levy 9 comet—smashed into the earth in 536 CE? To find out the answer, scientists turned to one of their favorite tools—computer models. When something happens only very rarely, it is very difficult to find enough instances in nature to fully understand it. A good example is the occurrence of ball lightning, a phenomenon that scientists are still struggling to comprehend. But with a mathematical model, we can create our ideal laboratory world inside a computer and then see the results when we cause various events—such as a small comet impact—to happen.

Astronomers have modeled the effects of a small comet impact on the earth.[16] They discovered that a small comet basically drills a hole through the atmosphere. As the comet explodes in the atmosphere, the debris is shot back through that air tunnel (much like a gun barrel), creating a plume of material that then settles back into the atmosphere. This model was used to explain the infamous Tunguska, Russia, explosion of 1908 that devastated a large part of Siberia. But astronomers theorize that the Tunguska comet—since it didn't lead to massive global cooling—was probably pretty small, maybe ten to twenty yards in diameter. How big would a comet need to be to cause the darkness and cooling recorded around 536 CE?

By working backward through the model, astronomers determined that the size of a comet needed to produce the amount of dust to create the 536 CE cooling was only six hundred meters—a bit bigger than six football fields. And using probability theory, they determined that the likelihood of an impact of such a comet is once every few thousand years—definitely within the limits of recorded history.

Could a comet have been the cause of the terrible weather and subsequent famine and disease of 536 CE? Before we answer definitely, we need to recognize that one of the best features of climate science is that it is constantly being updated. What we believed yesterday may not be our viewpoint tomorrow. As new evidence becomes available, different interpretations of the event become possible. For this mystery, a new analysis of the Greenland ice core data in 2008 caused Baillie to retreat from his theory of a comet strike causing the 536 CE solar dimming.[17] Ice core scientists carefully analyzed new Greenland ice core data and discovered indications in the ice record of a large volcanic eruption that they could date to 534 CE.[18] In particular, the ice core investigation revealed very high concentrations of sulfur, which is a key ingredient to global cooling. Consequently, they suggested that a high-sulfur volcanic eruption in 534 CE would be extensive enough to account for the solar dimming and therefore such a finding removed "the need for an extraterrestrial explanation."[19]

So we're left with plenty of tantalizing—but still changing—evidence. Something bad happened around 536 CE. It caused short-term (a decade or less) cold conditions as evidenced by the tree ring record and by the historical record. Some scientists suggested that a rare occurrence—in this case, a relatively small comet impact—might explain the event. Conversely, reinterpretation of the ice core record has reopened the possibility of a massive, sulfur-rich volcanic eruption to explain the event. No matter which theory is favored at the moment, it is likely that more scientific evidence will continue to be uncovered and it will provide new fuel for the debate.

So to close this mystery, let's give a final historical perspective of the earth's sudden weather change in 536 CE by the writer Baronius, as told by Dr. Noah Webster. Webster wrote:

> **The account which Baronius gives of this famine, is, perhaps more philosophical and deserves notice. He says, the crops failed, corn ripened prematurely, and was thin; in some places, it was not harvested, and that which was gathered, was deficient in nourishment. Those who subsisted upon it became pale, and were afflicted with bile. The body lost its heat and vigor, the skin dried, the countenance stupid, distorted and ghastly, the liver turned black. Many perished by hunger; many betook themselves to the field to feed on vegetables, and being too feeble to pull them, lay down and gnawed them off with their teeth.[20]**

Chapter 10
THE MYSTERY OF THE MAYAN MEGADROUGHT

Often, we tend to ignore the early history (and environment) of the Western Hemisphere—North and South Americas—for the much better researched history of Europe (and to a lesser degree, Asia and Africa). Yet the Americas have been home to an amazing set of sophisticated civilizations. What we are now discovering through the many sciences involved with climate change is that these civilizations were strongly influenced by their environments and occasionally by drastic changes in those environments.

In particular, Central America holds the same place in the history of the Americas as Africa, the Near East, and Europe do for the Egyptian and the Harappan civilizations. Civilizations arose in Central America as long as five thousand years ago. Over the course of time, a number of advanced cultures developed, matured, and declined in this region, including the Teotihucan, the Maya, the Zapotec, and the Mixtec, among others.

Time: 753 CE
Location: Tikel, located in the Yucatan Peninsula of present-day Mexico, North America

"I thirst! Please, daughter, bring me water."

Quaixalot was close to death. Her daughter made that painful admission as she gazed down with tear-filled eyes at her ailing mother. Yet she found little comfort in the knowledge that her mother's suffering would soon end. Her mother's illness was merely one of a seemingly endless number of such sicknesses throughout the city. While her mother was dying, thousands had already preceded her into death's bitter darkness.

Her mother's bedding was already soaked with sweat. A faint cloying smell of illness hung in the still, dry air.

"Peace, mother," her daughter cooed softly to the restless and fevered woman. Quaixalot's skin was a pale, yellowish color, and her daughter noted with increasing alarm that the large, hard nodules behind her ears were even larger than they had been the previous night.

Quaixalot's eyes suddenly opened wide—and apparently unseeingly—as she screamed in a wail of pain, clutching her chest.

"Please, mother. Rest," the daughter again whispered softly. She reached for a small pottery cup, filled with some of their precious water, and set it to the older woman's lips.

As the mother began to sip at the cup's liquid, the daughter looked up to see a thin, painfully emaciated boy of six enter the room. Her son—the older woman's grandson—looked at the two women silently.

"Was there any food available at the temple?" the daughter asked her

son quietly. She could see the answer already. The boy returned with empty arms.

"No, mother," the boy replied sadly. "They say everything has been eaten."

The daughter shook her head at the unfairness of the gods as tears continued to flow down her face. Her mother was soon to die—and her young son would likely soon join her. And there was nothing that the daughter could do. The helplessness abruptly reached a breaking point and the young woman began to cry in low, heaving sobs.

A bony rat, scurrying through the house in a desperate search for food, glanced up in momentary alarm at the low sobs originating from the weeping woman and then returned to its futile hunt for a morsel.

One of the central threads binding any civilization together is the availability of water. We saw with the Mystery of Petra that the Nabataeans used remarkable engineering to store and manage water in an environment that couldn't naturally sustain such a population. One aspect of climate variability is that, occasionally, change happens at a time in the society's cultural development such that the population, by either chance or design, is unable to react fast enough to the change. While that can be cruel enough, nature and humans in concert can sometimes compound the fundamental change by introducing related problems. In this mystery, we delve into how a climate change in Mesoamerica was aided by human development in creating a social catastrophe.

The so-called Classic period of Mesoamerican civilization, which existed from about 250 CE to 900 CE, produced incredible feats of engineering—for example, the creation of the famous stair-stepped pyramids found in the Yucatan Peninsula of Mexico and other Central American locations. The period is also marked by the growth of large urban city-state complexes made up of civilizations such as the Mayan. The Maya achieved notable intellectual feats, including the development of a very accurate calendar system, the creation of the graceful artistic carvings at places such as Palenque and Copán, the growth of a remarkably advanced set of astronomical observations, and the intriguing systems of writing and mathematics.

Climatically, the Mayan civilization developed in a region where the rainfall is very seasonal. Its summertime had a markedly wet period but there was an equally pronounced dry period in the winter. Because of this great seasonal disparity in available water, the Mayans developed amazingly sophisticated methods for moving and storing water. For example, after quarrying the rock and stone used to make their beautiful buildings and pyramids, the quarries were often converted into water reservoirs. In addition, they developed an elaborate system of complex hydroengineering similar to the Nabataeans of Petra, composed of miles of canals and unique urban water management facilities including dams, floodgates, and desilting tanks. This system created a safety net to the emerging urban centers. For example, archaeologists have estimated that the water facilities of one city might have provided stored water for nearly ten thousand people for six months if the seasonal rains failed.

This sophisticated civilization flourished until about 750 CE. Then, within the short space of a century or so, the entire Mayan population of the Central American lowlands was decimated. There are no hieroglyphic dates anywhere in these regions after 910 CE, not until the rise of the Mayapan society in the late twelfth century.

The critical mystery is simply what caused the utter collapse of this vibrant civilization?

Let's first look at the climate of that time. This time we'll use a different way to employ one of the weapons in our climate analysis arsenal, sediment cores. In the Mystery of the Vanishing Harappans, I discussed how palynologists—scientists working with pollen variations over time—analyzed their pollen data from sediment cores, which are drilled tube samples taken from lake-bottom mud, to determine what type of plants and crops were grown in a given area. From those pollen samples, the palynologists could extrapolate the environmental conditions of that time.

Another way to examine a sediment core is to study the chemical makeup of the layered mud samples themselves. Researchers from the Department of Geological Studies at the University of Florida carefully extracted sediment core samples from lakes in the Yucatan Peninsula, the central region of the Maya.[1] Those scientists discovered that the sediment cores displayed significant variations in the amount of gypsum throughout their depth.

That was important because the gypsum concentrations in the

lake-bottom mud are significantly linked to the level of the lake, which in turn is a function of the annual rainfall. When lake levels are high, the highest concentrations of gypsum are found in the mud of the shallow water at the edge of the lake with much smaller gypsum amounts in the deeper water muds of the lake. But under drought conditions when lake levels are low, the gypsum is found equally dispersed throughout all of the available water areas of the lake. Consequently, if the center of the lake is drilled, layers of the sediment rich in gypsum indicate past drought conditions while sediment layers low in gypsum concentration suggest past periods of wetter climates. When those individual layers are dated through isotope analysis, we can establish for the region in which the core was drilled exactly what times in the past were wet and what times were dry.

So, by studying the gypsum concentrations in lakes in the Mexican Yucatan Peninsula, the central region of the Maya, the University of Florida researchers determined that there were three different extended dry periods: namely, droughts in 475–250 BCE, 125–210 CE, and 750–1025 CE. Does that prove the Mayan civilization died off solely due to drought? The answer might not be so clear-cut. An interesting study based on dendrochronology has suggested that the Mayan collapse may have been caused by more than just drought. To fully interpret that research, let's continue our survey of the science of dendrochronology.

As I mentioned in the last mystery, dendrochronology is very young—it was initially established by A. E. Douglass only at the beginning of the last century. He had determined that the more "stressed" members of certain types of trees displayed variations in their ring widths that corresponded to climate changes. By linking the overlapping portions of the tree ring records of living and dead trees (the process that dendrochronologists call cross-dating), very long databases of reconstructed climate could be created.

However, these early dendrochronology studies simply assumed a direct and straightforward environmental control of tree growth; they held that the changes in tree rings were directly related to the coinciding climate. For example, a big tree ring would be interpreted as indicating that good weather occurred during that growing season. But that is, unfortunately, far too simple. Regrettably, a tree is not a finely tuned biological thermometer or a living rain gauge. In other words, tree growth is not necessarily showing a direct and precise response to climate. Other factors such as soil nutrients, species competition, and even fire or predatory animals can also influence tree growth in a given year.

Since the physiological processes between tree growth and climate linkages are too poorly understood (and perhaps too complex) to allow the creation of mathematical equations of plant/climate interaction—we don't have "tree-weather computer models"—dendrochronologists must resort to less powerful techniques based on statistical analyses of actual plant-climate observations. One of the most common of these is called the *response function.*[2]

In developing a response function, we attempt to explain or predict a given year's ring width as an autoregressive function of prior growth, and as a lagged and direct response to a multitude of climate factors. Ouch—let's take that simply. *Autoregressive* means that, to some extent, a tree's growth during one year is dependent on growth from previous years. In the same fashion, the weather of the current month or season (direct response) and the previous months or seasons (lagged response) can also exert a major influence on a given year's tree growth. Statistically, we can extract just the critical common information from all of these potential influencing factors, that is, the autoregressive, the lagged, and the direct factors. The statistical technique that reduces all of these factors down to a much smaller set of common variables is called *principal components analysis*, or PCA.

Let's illustrate the concept using a simple example of PCA. Suppose we have a data set of annual modern tree ring variations and the actual records of monthly temperatures and precipitation totals for the area. First, we collect all of the similar climate information (for example, the temperatures and precipitation totals for all of the months that influence a given year's tree ring growth). If we used all climate information over the previous, say, two years, we would have at least forty-eight potential variables (twenty-four monthly temperatures and twenty-four monthly precipitation variables). With that many variables, we risk the chance of what is called *multicollinearity*, using two or more variables to explain the same thing. Multicollinearity creates the danger of two or more monthly temperatures equally explaining a certain portion of the tree ring variations. If we continue to use both monthly temperatures, we could run into future problems with the reliability of the statistics. Consequently, the better method is to extract just the specific shared information in all of the monthly temperatures and use only that information. Principal component analysis does exactly that—it extracts the core information that is most shared between all of the factors (the autoregressive, the lagged, and the direct). So rather than having to work with, perhaps, forty-eight potential factors, we might only have three or four shared variables, or what are called *eigenvectors*.

Once we have those three or four factors, we need to see exactly how they influence tree ring variations. To do that, we employ another statistical technique—*regression analysis*. Regression analysis relates the variations in one variable, called the dependent factor (in this case, the tree ring variations), with the now-much-smaller set of independent factors (in this case, the PCA-modified set of climate variables). Regression determines how well the independent factors can explain (or predict) the variations in the dependent variable. The end result is a mathematical equation called the response function, which computes the variations in tree ring width by using the set of PCA climate variables.

Next we have to determine how well our response function works. At this point, we have tested the response function only for the calibration period (the period that contains both the tree ring data and the measured, comparative climate data). That doesn't guarantee a successful reconstruction for a different time period. To better establish the reliability of a calibration, an independent verification is necessary. Such verification can be accomplished by three different methods: statistical verification with independent meteorological data (such as nearby temperatures or precipitation values not used to create the response function), verification with independent proxy data (for example, pollen analysis as discussed earlier), and/or verification with historical records (such as written documents about the weather).

But the response function isn't what we want for historical climate research. We want the tree rings to predict climate change, not the other way around! We need to have the tree rings tell us the climate for times and/or places where there was no other climate information. So, we use the response function to create a *transfer function*, a statistical relationship that converts the tree ring measurements back to climate variations.[3] With that relationship—and assuming that tree rings for past times have responded to climate in the same manner that they are responding today—we can reconstruct maps of climate (normally temperature and precipitation) for periods long ago.

For this mystery in particular, we can compute the climates for the Americas far back before we have written records of rainfall and temperature based on the response and transfer functions. So how do the dendrochronologists interpret the reconstructed climate record of the region for the time of the Mayan civilization's collapse? Based in part on an intriguing analogue with the sixteenth century, I can answer that question with one word coined by a past brilliant student of mine, Professor Dave Stahle from the University of Arkansas: *megadrought*.

Most droughts are relatively short-term events that last only a few years. Sometimes they can last longer—much longer.

Stahle and other dendrochronologists first examined the rainfall record that they reconstructed from their tree ring transfer and response functions for Mexico during the sixteenth century.[4] That tree ring reconstruction of rainfall, which matched well with other records of regional rainfall, indicated that a very severe—and particularly long-lasting—drought occurred in Mexico and the Southwest United States during most of the 1500s. They termed this abnormally long dry period a megadrought.

Recall that the University of Florida researchers determined that for the Mayan civilization, three different extended dry periods could be seen in the gypsum variations in their sediment cores: namely, droughts in 475–250 BCE, 125–210 CE, and 750–1025 CE.[5] It is that last long drought that bears close examination in that its length and extent resembles features of the megadrought that Stahle and his colleagues found via dendrochronology for the sixteenth century. However, one aspect of the Mayan eighth-century megadrought that was different from its sixteenth-century counterpart was its more periodic nature; it was actually more of an extended series of droughts separated by brief, but intense, wet periods. Could such a megadrought have caused the fall of the Mayan civilization?

Archaeologists have determined that the Maya and others in the region had developed a rather impressive system of hydroengineering to collect, store, and transfer water so that periodic droughts wouldn't have catastrophic impact. These periodic pulses of the eighth-century megadrought took place during the height of several of the most advanced pre-Columbian civilizations. Couldn't some of their water engineering techniques, similar to those of the Nabataeans at Petra, allow them to survive until better conditions arrived? In this case, the answer is apparently no.

Stahle and his colleagues had first defined megadrought—meaning a very severe and particularly long-lasting drought—for the 1500s, the time when the Mesoamericans first came into contact with Europeans. At the time of first contact in 1519, the population of Mesoamerica was estimated to be perhaps as large as twenty-two million people. Within a single short century, that population had collapsed to only two million people—over 90 percent of the entire population had died. Part of that collapse was due to the importation of diseases, such as smallpox, from Europe. However, researchers now suggest that the vast majority of the deaths resulted from a hemorrhagic fever called *cocoliztli*, a type unknown to both Mesoamerican and European peoples.[6] It was characterized by symptoms of a high fever, severe headache, great thirst, and weak pulse, coupled with jaundice. After infection, the victims of *cocoliztli* became restless and frenzied, and then developed hard, painful nodules behind the ears. These conditions eventually led to stomach pain and dysentery, eventually terminating in a bleed-out from the ears, mouth, and nose. Unfortunately, even for the few initial survivors, well-being was not guaranteed. The survivors were subject to relapses—and were often in such a weakened condition that relapses led to death.

It is important to emphasize that the sixteenth-century megadrought didn't cause *cocoliztli*—but the aridity aided its spread and virulence. Specifically, Stahle and his colleagues discovered that the *cocoliztli* epidemics occurred during a couple of brief but intense wet periods that occurred during the megadrought. They noted that a specific drought/rain pattern—long sustained dryness, then a quick transition to wet, eventually returning back to aridity—is often the perfect climate for sudden expansions of an animal disease–host species. For example, the deadly hantavirus propagated throughout the Southwest North America through the sudden proliferation of deer mice during the short-lived wet conditions of the early 1990s.[7] Stahle and his colleagues reasoned that a similar situation likely happened in the sixteenth century to create the conditions necessary for the pandemic.

Continuing this line of logical reasoning, Stahle and his fellow researchers concluded that the more sporadic megadrought of the eighth century created even more favorable conditions for the development and eventual dispersal of the *cocoliztli* hemorrhagic fever than did the megadrought of the sixteenth century.[8] By punctuating a long period of below-normal rainfall with a few wet torrents, the killer disease was likely able to resurface to infect a new unsuspecting—and increasingly helpless—population. With dwindling food supplies and an ever-decreasing population base, the advanced civilizations of the Classic Mesoamerican period finally succumbed to the dual forces of drought and disease.

Now, bear in mind, this is a deductive line of climatological reasoning. Many other possible explanations for the collapse have been postulated. War, social unrest, and migration have also been offered as alternative reasons for the collapse of the Mesoamerican civilizations. And—as Stahle and his colleagues are quick to point out—it is possible that multiple causes might have been operating.

Yet the beauty of an environmental theory such as the megadrought/disease scenario is that it is potentially testable. As we gain new facts and understandings of the environmental conditions at the end of the pre-Columbian Classic period of history, we can modify, expand, or even abandon elements of this theory to better match the new information.

In this case, the development of new technologies of scanning human remains might eventually identify how the hemorrhagic fever of the eighth century was spread. We might ultimately determine if an animal host population—such as the hantavirus's deer mice—was responsible for the *cocoliztli* pandemic through population explosions linked to varying drought/flood cycles.

So what is the message of this rather morbid mystery? I'm afraid it isn't a happy one. Dave Stahle and his fellow researchers phrased it best when they stated that perhaps the most important aspect of their study was to encourage new research on "whether there may be a highly lethal microorganism still present in the climate-sensitive animal reservoir in Mesoamerica."[9] In other words, they were asking "Could this type of disaster happen again?" Perhaps that doesn't scare you, but it sure scares me!

Time: 754 CE
Location: Tikel, located in the Yucatan Peninsula of present-day Mexico, North America

Ziolotox stared down at the empty bed of her beloved mother. Last year, Quaixalot had finally succumbed to the terrible ravages of the hideous disease that was driving her people ever closer to extinction. The old woman had been among the first to die from the virulent sickness . . . but she had not been the last to die.

The dreadful malady spread. Soon Ziolotox's young boy had caught the terrible illness and—perhaps fortunately—died not long after. She shook her head sorrowfully. Her young boy and her mother had been but two of the many who had suffered the terrible signs of the illness—the fever, the swellings . . . and the final death.

She glanced out the doorway at the clear, blue sky. As if mocking her, the sky was solemnly peaceful—only a few delicate puffs of cloud dotted the heavens.

It was all so unfair! First, her people had endured that horrible famine for years. It had seemed as if the skies refused to give them the life-giving waters for their crops—but they had been able to cope with the dryness. Irrigation, trade—those things had worked for a while. But then, last year, when it had finally appeared that the gods had at long last favored them once again with bountiful rains, the pestilence had spread among the people . . . and killed so many.

Ziolotox heaved a heavy sigh. So many had died that her store of tears had been drained.

And now, she thought as she looked back up at those maddeningly dry blue skies, the gods had once again resolved to withhold the precious rains from her people. They no longer had the manpower to work and maintain the irrigation canals, to send traders north and south, or to even handle the multitudes who were still sick.

Ziolotox turned back to her small house. The life of

her childhood—the beautiful festivals, the magnificently colored gowns, and the rich foods—was gone forever. Today, simple survival was the most that she and her people could ask of the gods.

Chapter 11

THE MYSTERY OF COLUMBUS'S MISSING HURRICANE

Sometimes the mystery is when the dog *doesn't* bark. I can illustrate this point using the 1892 Sherlock Holmes story "Silver Blaze." In that story, a client responded to Holmes's comment of "the curious incident of the dog in the night-time" by noting in puzzlement that the dog had done nothing during the night. The illustrious detective replied, "That was the curious incident."[1] In this weather mystery, like in the Holmes story, our puzzle doesn't involve the presence of something—rather, we need to explain its absence.

All schoolchildren know from the old nursery rhythm that "in fourteen hundred and ninety-two, Columbus sailed the ocean blue." Yet, what many schoolchildren—and even adults—don't realize is that Christopher Columbus did not encounter any adverse weather when he sailed his three small galleons across the Atlantic Ocean. As recorded in his journal, the thirty-six-day voyage of the *Niña*, the *Pinta*, and the *Santa Maria* across the ocean blue was exceedingly dull—at least with regard to the weather. It was quite definitely *not* the stuff of modern blockbuster movies. No ravaging hurricanes, no ripping winds, not even much rain. The question may be raised: Is that surprising?

Time: October 2, 1492
Location: Atlantic Ocean
(This and the closing vignette of this chapter are loosely based in part on entries recorded in the purported logbook of Christopher Columbus.)

The admiral of the fleet sat behind his pinewood desk, in the midst of logging the day's events in his journal. He sat with the ease of an experienced sailor, not even noticing the gentle swaying of his ship as it rocked in the low swells of the placid ocean. The oil lamp above his desk swung in a gentle arc back and forth to the rhythm of the mild waves, causing the flickering light to play across the pages of his journal.

He dipped the goose quill into a small glass inkwell with the habitual care of a meticulous chronicler and then paused as he considered the words that he would write in his journal. The day's travel had been exceedingly good—from the pilot's numbers, he had computed a total day's travel distance of 117 miles. Of course, to allay the growing fears of his superstitious crew, he had proclaimed a distance of only 90 miles to the sailors.

The thought caused him to shake his head in frustration. Confound it all, these rising qualms of the crew at being so far from sight of land were becoming evermore worrisome. Today, some of the crew had become unduly alarmed at a bit of seaweed seen drifting past the fleet from the east—literally from behind the ships! They had begun to clamor that we had traveled too far. "Land lay astern, not ahead" was the growing whisper among the crew.

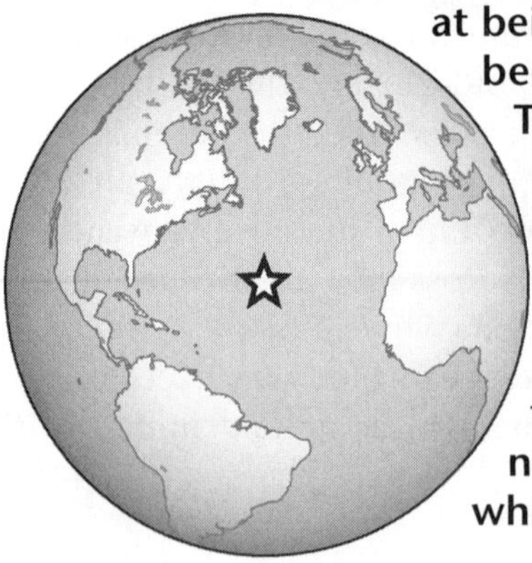

The admiral sighed. The responsibility of the fleet, the expectations of his benefactors, and the realities of travel were all increasingly wearing on him. Too many days of these so-called omens and there would undoubtedly be trouble with the crew.

Still, it could be worse, much worse, he reflected, as he tried to focus upon the positives. The sky, once again, had been beautifully clear and there had again been a good easterly wind to push his fleet forward. The sea continued to display a benign face with only gentle swells. The waters were even bountiful—today one of the crew had managed to hook a goodly sized fish.

Indeed, the admiral realized, the expedition could have turned ruinous had a storm struck the fleet. Being this far from Spain—or even the Canary Islands—did not allow for any errors in the functioning of the small fleet. The benign weather, despite allowing idleness and grumbling to foster among the crew, was much more of a blessing than a curse.

Yes, the admiral decided. Discipline—and his continued deception about travel numbers—still was the best course of action to handle his querulous crew.

Confidently, he pulled the quill from the inkwell, then resolutely began to pen.

"The sea is smooth and favorable. Many thanks be given to God . . ."

Hurricanes have long been considered to be the scourge of sailors throughout the Caribbean Islands. In fact, during the days of colonial exploration, the Roman Catholic Church decreed that a special prayer, "*Ad repellendam tempestatis*," should be recited during the hurricane season to protect the faithful from the ravages of those monster storms. Yet, despite such prayers, thousands of ships have suffered disastrous encounters with hurricanes. The Great Hurricane of 1780, for example, sank at least 1,200 ships and killed more than 22,000 people. In the Atlantic Ocean, the months of highest occurrence for these killer storms are August through October, with September being the month with the highest incidence of hurricanes.

Admiral Christopher Columbus left the Grand Canary Islands on the sixth of September and arrived in the New World on the twelfth of October 1492. Consequently, he traveled the entire width of the Atlantic Ocean during the height of the hurricane season. Why did he not only fail to encounter any hurricanes but not even experience any significant weather of consequence?

A premier historical climatologist, David Ludlum, wrote, "The outstanding meteorological fact of the First Voyage is simply that no hurricanes or severe storms were encountered in the West Indies despite the fact that the fleet of small vessels traversed an area of tropical storm activity at the season of their most frequent occurrence."[2] The implication is that Columbus would have considered himself very lucky not to encounter an Atlantic hurricane had he known the likelihood of storms in the area. But is this true? Was Columbus lucky in that the weather of 1492 was less prone to hurricanes or was he simply a superb navigator who chose a route that was the most favorable to crossing the Atlantic? Dr. Jay Hobgood, a hurricane climatologist at Ohio State University, and I undertook an investigation of that intriguing mystery.[3]

For this mystery the weather detective's tool of choice is the historical record. Columbus and his pilots recorded a variety of information into their ships' logs during the thirty-six-day voyage to the New World. These included notations on the color of the water, the presence of certain types of birds and fish, as well as impressions of winds, currents, and weather. From Columbus's log, a total of sixteen separate references were made about the wind.[4] In general, these notations related to the strength of the winds. Eleven days of calm or near-calm conditions were cited in Columbus's journal, four days of "mild" or "soft" breezes, and one occurrence of "freshened" winds.

We can gain valuable information by comparing the distance that Columbus traveled on "calm" days to the distances sailed during the rest of the voyage. Using distances cited from the logs (as privately computed by Columbus, not the values he gave the crew to allay their fears of open-sea travel), the mean distance traveled on days labeled "calm" was fifteen leagues (but I should note that there is considerable disagreement among historians as to the exact distance of a Columbian league). Conversely, the distance traveled on all other days was thirty-nine leagues. While the measurements undoubtedly contain some error (after all, they were made on a moving vessel in a moving ocean), such a difference demonstrates sufficient data quality to allow comparisons between the distances traveled under different wind conditions.

If we accept that the Columbian pilots were accurate in their description of "calms"—and assume that the relativistic movement of the ships and other effects were negligible—then the fifteen leagues traveled under those "calm" conditions should have been caused primarily by the speed of the ocean current. Additionally, if we assume that a Columbian league is 4,286 English feet as some historians have suggested, we can then calculate the speed of travel by Columbus under "calm" conditions—propelled just by ocean currents—to be 0.42 knots. How does that compare to modern values? Modern estimates of current speeds for this portion of the Columbus track give an average speed of 0.4 knots. A very favorable comparison! That suggests the Columbian data are indeed of good quality.

As mentioned above, the Columbian ships experienced calm or near-calm winds on eleven out of thirty-six days, or about 30 percent of the time. Using modern sailing charts from the last hundred years, Hobgood and I found that the average modern frequency of calms in that region of the Atlantic is between 3 and 15 percent. So, even with the most liberal estimates, Columbus likely experienced more than

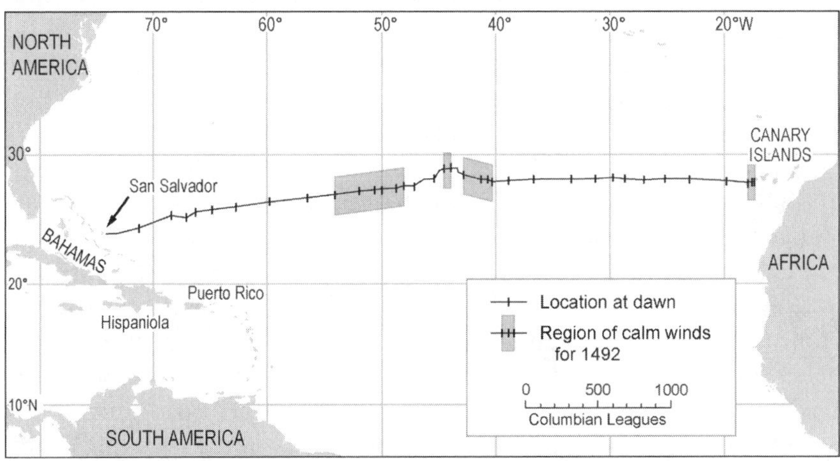

Map showing the probable route of the three ships of Christopher Columbus with regions of calm winds noted. Map by Barbara Trapido-Lurie (adapted from Ref. 3).

twice the number of calms than occur in the region presently. Is that climate change? Before we decide, let's look at one more aspect of calms that we can address from the Columbus logbook.

If we plot the positions of where the three Spanish galleons experienced those calm conditions and compare those positions to where calms occur today, we can see if climate change has indeed taken place. Columbus traveled a fairly consistent westerly route along 25–27°N. Modern wind charts for the Atlantic show that track to be strongly impacted today by calm conditions. Consequently, given that Columbus experienced twice as many calms as experienced today in the same region, we can suggest that the high number of calms in 1492 was not the result of any north or south movement of high pressure (the type of weather system that tends to create calms). Rather, it was an intensification of an existing high-pressure system normally found (both today and likely in 1492) around 25–27°N.

From an east/west viewpoint, Columbus encountered three distinct regions of calms, one near the Canary Islands, a second set between 40°W and 45°W, and a third set between 48°W and 52°W. Those locations match locations that today tend to experience calms.

So Columbus chose a path across the Atlantic Ocean that did—and does today—experience significant calm conditions. In fact, the year 1492 appears to have been an exceptionally (at least compared to modern day) calm year. Does that meteorological quietude extend to hurricanes as well?

Once again, Hobgood and I turned to the historical records. In 1987, two oceanographers (Roger Goldsmith and Philip Richardson) produced a computer reconstruction of Columbus's most-probable trek across the Atlantic Ocean.[5] Their computed landfall was a small island called Watling Island, which is considered to be one of the possible landing points of Columbus. Using that computer reconstruction, we compared the dates and locations of Columbus's fleet against the date and position of every known tropical storm and hurricane recorded over the last century. Surprisingly, given the hundreds of tropical cyclones that have been spawned in the Atlantic Ocean over the last century, only four storms actually crossed the route of Columbus—three between 1900 and 1906 and a fourth in 1949. Of these four, only one actually reached hurricane status.

Therefore, if modern frequencies of hurricanes correspond to those of 1492, the chances of the Columbian fleet receiving a direct hit from a tropical cyclone would have been extremely small—only 4 chances in 140 of record. If we restrict our investigation to only storms of hurricane intensity, a hurricane has only intersected Columbus's path less than once a century.

So what were our conclusions? Was Columbus lucky or was the weather different in 1492? While Columbus may have been climatologically fortunate to have avoided an encounter with a hurricane, the odds were decidedly in his favor. Columbus's famed weather luck was most probably a combination of two factors. First, he likely encountered stronger-than-normal high pressure over the North Atlantic (whether this is climate change or just normal year-to-year variability is impossible to say). Second, he plotted a sufficiently northerly voyage as to have avoided the region that commonly experiences Atlantic hurricanes. Fundamentally, our analysis suggested that Christopher Columbus was an exceedingly good sailor.

Time: October 12, 1492
Location: A small Caribbean island (perhaps Samana Cay or Watling Island), Atlantic Ocean

Today had been the culmination of much work—much good work! With the final wave to the natives departing the *Santa Maria,* the admiral of the fleet strode to his cabin and then walked purposefully to his pinewood desk. As he sat down, he opened the leather-bound logbook to record the events. The ship—now safely anchored in the small harbor of this quiet island—no longer swayed as it had across the great expanse of ocean that they had so laboriously sailed over the past month. He stood up to adjust the flame on the oil lamp hanging limply above the desk, and then sat down to write in his journal.

It had indeed been an eventful day. At dawn, the admiral—together with the captains of the *Pinta* and the *Niña*—had put to shore in one of the longboats. Excitement had been clearly evident in the face of every sailor. Not only had they reached land, but it was obvious that this island was inhabited—many naked

people stood watching the galleons from the shore. But the natives had not interfered as the admiral had unfurled the royal banner and gave a short prayer of thanksgiving for the safe landfall.

Despite the complaints of his crew, the voyage had been safe and uneventful.

As he carefully wrote that information into the log, the admiral paused. There was one more critical point that he needed to make in the log—something to which future explorers to this area should take solemn notice. He reflected on the kind demeanor of the natives. Yes, the key to this whole expedition was friendship, friendship between the Europeans and the natives—*we should treat the natives with respect . . .*

He picked up the quill.

"I want the natives to develop a friendly attitude toward us because I know that they are a people who can be made free and converted to our Holy Faith more by love than by force . . ."

Chapter 12

THE MYSTERY OF THE LITTLE ICE AGE'S LOST SUNSPOTS

Over the course of the last five hundred years, Western civilization has endured markedly colder temperatures than presently exist. Indeed, the time period from 1500 to 1850 is called the "Little Ice Age" because of the consistently frigid temperatures that hovered over Europe and eastern North America. For example, during those cold centuries, over half of the population of Bohemia (the modern Czech Republic) died of starvation.

Time: January 14, 1620
Location: London, England, Europe

Alyssa was having a hard time keeping up with the drink orders. It was a merry night in the warm tavern—everyone was quite happy not to be out in the bitterly cold darkness. And, Alyssa noted with a smile, all of the tavern patrons were celebrating that joyful fact by

drinking considerable quantities of beer. A light snowfall had blanketed the city, covering the dingy gray buildings with a glittering coat of white.

"How cold was it? I'll tell ya how bloody cold it was!" one voice rose above the babble of normal pub talk.

Oh, dear Lord, Alyssa thought. Now they be tellin' the "cold" stories again. With so many recent years of bitterly cold winters, a growing pastime at many of the taverns on many a frigid evening was the telling of "cold stories" . . .

"Me grandfather, he told me when I was but a wee tot that one frigid winter it was so cold that when a hard frost hit followin' a winter flood, a solid swarm of frozen flies and beetles flowed down the Severn River for days! He said that the mills were dammed up with the frozen bugs for four days, and that the men had to clean them out with shovels."

"Oh, yeah?" another patron yelled out with a laugh. "Well, me grand'fer once said that a winter long ago had such deep snow that the pallbearers at a funeral had to carry the coffin over a field gate rather than through it! And the mourners had to climb over the hedges!"

"Well, I don't know about that," another man exclaimed. "But I do know that the Ouse once froze hard enough to hold a horse-race from the tower of St. Mary Gate to the Crain at Skeldergate postern!"

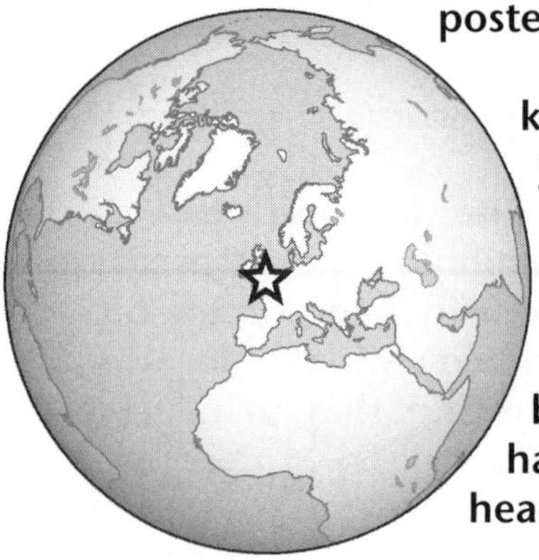

A grizzled seaman spoke up, "Ye know nothing of cold, I be tellin' ya. I was in the New World to the West, in Jamestown itself, and I be knowin' something of cold. The gov'nor there, he said in the bitter winter of '09 that anybody be caught stealin' food would have his ears sheared right off his head! Even still some of the starvin'

people began to eat their leather clothing. I hears tell that some folk even dug up winterin' snakes and chewed 'em up raw!" The grizzled sailor lowered his voice, "Indeed, some men, I heard tell, did dig up corpses from their graves for the food!"

A load roar of disbelief gave the general reaction from the tavern's clientele, but Alyssa shivered in sympathetic reaction and glanced at the cheery warmth of the huge hearth. It was a good night to be inside next to a warm fire!

In recent times—I'm talking of the last five hundred years or so—it has been cold much more often than it has been warm. The Little Ice Age from 1500 to 1850 saw unswervingly freezing temperatures over Europe and eastern North America. One very visual aspect that historians tell us of that time of bitter cold was the occurrence of "frost fairs" on the river Thames in London. Quite simply, the Thames doesn't freeze today. In part this is due to a general increase in global temperatures as the region has recovered from the Little Ice Age, in part due to an increase in river flow due to the demolition of the old narrow London Bridge, and in part due to the creation of power plants that have put warmer waters into the river.

However, when the Thames did freeze in the past, the event often became a time of a great festival. Merchants would open shops on the ice using booths they sledded onto the ice and the frozen river would be a great beehive of commerce and entertainment. One of the first frozen Thames entertainments was soccer. Raphael Holinshed recorded a frost in 1564, "which continued so extremely that on New Years Eve, people went over and along the Thames on the ice from London bridge to Westminster. Some played at the football as boldly there as if it had been on the dry land."[1] Even Queen Elizabeth crossed the frozen river during that frigid winter.

By 1607, the ice festivities had grown. The Water-millock *Register* recorded "a marvelous great frost which continued from the first day of December until the 15th day of February after. . . . So strange that men in great companies made common way up the same . . . with horses loaded with corn. Upon the 6th day of January the young folk of Sowlby went unto the midst of the [frozen river] and had a Minstrel with them and there danced all the afternoon."[2]

The frost fair in 1684 had become much more elaborate. John Evelyn described the situation: "I went across the Thames on the ice, which now became so thick as to bear not only streets of booths, in which they roasted meat, and had diverse shops of wares, as in a town. . . . The Thames before London was planned with booths in formal streets, all sorts of trades and shops furnished and full of commodities, even to a printing-press, where the people and ladies took a fancy to have their names printed on the Thames."[3]

The last of the great Thames frost fairs occurred in 1814. According to one newspaper, this fair even included the unusual spectacle of an elephant being walked onto the ice.[4] Perhaps a bit of exaggeration has crept into these reports, for the reputable *London Times* reported, "We observed several booths erected upon [the ice-filled Thames] for the sale of small wares; but the publicans and spirit-dealers were most in the receipt of custom."[5]

But beyond the festivities, people also died. The horrible winter of 1609 almost destroyed the colonists of the Jamestown colony in Virginia. By December of that year, the daily ration of grain was reduced to half a pint of corn and, according to one account, each can of corn "contained as many worms as grains."[6] It is said that starving men sometimes even bartered their clothes for food—but then, of course, they froze all the more quickly. And the cold wasn't limited to the New World. The Reverend Thomas Short in his fascinating 1749 book, *A General Chronological History of the Air, Weather, Seasons, Meteors, etc. in Sundary Places and Different Times*, wrote that during 1665 in Hungary, because of the cold, "People, to preserve their Lives, were forced to get into Ovens heated with Wood."[7]

All of this bitter cold took place over a 350-year period—probably the coldest period human civilization has witnessed in the last 10,000 years.

The mystery is straightforward: What caused the Little Ice Age?

To solve this mystery, we must delve into the science of astronomy—and, in this case, the detailed historical observations of our sun.

The first critical solar fact in this mystery is that the sun has periodic bouts of pronounced "solar acne"—Earth-sized black blotches called sunspots that appear on its surface. These blotches are massive—as much as thirty thousand miles in diameter—but, as compared to the rest of the sun, they are relatively cool in temperature and short lived in duration. A typical sunspot lasts only a few days or weeks before it breaks up and disappears.

We have historical records of sunspot observations, particularly those made in China that extend back for two thousand years or more. Throughout history, many peoples—particularly the Chinese—have noted relationships between solar activity, climates, harvests, and the country's (and emperor's) good or bad fortune. These records were carefully preserved and used as a basis for astrological forecasting. Astrologers who failed in their predictions were routinely executed, so it is perhaps understandable that they diligently made daily solar observations. In this modern era, we should be aware that a person should not make observations of the sun using his naked eyes because of the high risk of blindness.

In contrast, Western records on sunspots are generally considered complete only from 1611, dating to the time when Galileo first noticed dark splotches on the solar surface after viewing the sun with his new telescope. Following that first Western discovery, other astronomers quickly began observing—and recording—the existence and the number of these large, dark solar regions on a regular basis. The Zurich Observatory in Switzerland initiated daily continuous records in 1749 and, with the addition of other observatories, a well-maintained continuous recording of sunspot numbers commenced in 1849.

The astronomers of these observatories soon developed rigorous procedures for consistent observation of sunspots. They defined—and computed—a "sunspot number," which they calculated as the sum of the number of individual sunspots and a value defined as ten times the number of clusters or groups of individual sunspots. Since most sunspot groups have on average about ten spots, this formula for counting sunspots gave reliable numbers even when the observing conditions were less than ideal and small spots were difficult to see.

Modern studies indicate that sunspots are regions in the solar photosphere that are marked by a darker hue, a much lower temperature ("only" 4,000°C, or 7,200°F, compared to 5,500°C, or 9,900°F), and a much more intense magnetic field than their surroundings. A sunspot's immense magnetic field often leads to the development of huge solar discharges, or flares, called solar storms.

The reason for such storms is that the intense magnetic field of a sunspot causes a massive plasma flow into or out of the sunspot (depending on the sunspot's polarity). A critical modern aspect of these solar storms is that they can cause major disruptions of orbital communication (which includes cell phones).

Although individual sunspots don't last long—only a few days to a month or so at most—the sequence of sunspot growth and decline follows a rather consistent near-decadal (ten-year) cycle. Rudolf Wolf, a director of the Zurich Observatory, was one of the first scientists to note that this sunspot cycle has an average period of about eleven years. There is also an abrupt reversal of the polarity of the sun's magnetic field so that a complete magnetic cycle (the so-called Hale cycle, or double-sunspot cycle) is completed in approximately twenty-two years. One final basic question: What causes these eleven-year and twenty-two-year cycles in the number of sunspots? Our best theories today suggest that they are linked to the gravitational influences of the solar system's planets. This is because some of the harmonic (longer-term periodic) variations in sunspot numbers correspond to the annual revolutions of the planets—particularly Jupiter and Saturn.

In modern times, we have also measured the relationship between the number of sunspots and the amount of energy emitted by the sun. Oddly enough, we find that a high number of dark sunspots generally corresponds to a greater amount of solar radiation being emitted by the sun. This may seem a bit strange since sunspots themselves are darker and cooler than their surroundings—and consequently, they emit less radiation. Yet the sun as a whole emits more radiation during high sunspot activity because, in addition to the sunspots themselves, intensely bright regions called *faculae*, or *plage*, surround these magnetic storms. And, importantly, when sunspots are at the edge of the sun from our vantage point (that is, just rotating onto or off of its face), the faculae create a significant increase in radiation that far outweighs the small dip in solar flux caused by the sunspot.

When Wolf began recording sunspot numbers from his observatory in Zurich in the nineteenth century—and when historians dug into long-forgotten written observations of the sun—we eventually archived longer and longer records of sunspot activity. It was through analysis of those long records of sunspot numbers that researchers began to notice something very odd. Plots of sunspot activity through time showed a very marked—and unexplained—absence of sunspot activity throughout the late seventeenth century (basically from 1645 to 1715).

The question that astronomers quickly asked themselves was simply: Is the pronounced lack of sunspots in the seventeenth century real or were people simply not observing the sun for sunspots during that time? The problem is that a "negative" is a notoriously difficult thing to prove in science (or even in a court of law). How do you show a "lack" of something? Luckily, an astronomer by the name of John Eddy did just that when he wrote a landmark article in the technical journal *Science* in 1976.[8] But actually, he didn't do it himself. As Eddy was quick to point out, he was simply reviving and extending the research of two earlier astronomers, Gustav Spörer and E. W. Maunder. In the nineteenth century, Spörer and Maunder had both tried to demonstrate that there had indeed been a "prolonged sunspot minimum" between 1645 and 1715. Unfortunately, other scientists almost universally ignored their investigations at that time.

So in a way Eddy was attempting to revive long-dead research. He began by combing through the old scientific journals. He even said that the task of uncovering whether sunspots occurred between 1645 and 1715 was "a detective story: a crime, serious for astronomy and perhaps for the Earth itself."[9] And what Eddy uncovered was forgotten detailed inquiries by another scientist. E. W. Maunder had actually been the original sunspot detective—an astronomer who had been ably assisted by his own Dr. Watson, his second wife, Annie Maunder, a Cambridge-trained mathematician.

It had been Maunder and his wife who first diligently investigated historical records to verify the lack of sunspots in the seventeenth century. During that search, the British couple had stumbled across a 1671 article in the *Philosophical Transactions of the Royal Society* that read in part, "At Paris the Excellent Signior Cassini hath lately detected again Spots in the Sun, of which none have been seen these many years that we know of."[10] In fact, the author of that 1671 letter then went on to review a sunspot description of eleven years previous for those who might have forgotten what a sunspot looked like. Another French astronomer, Jean Picard, observed at about the same time that he was pleased "to observed at Sea a Spot in the Sun from the third of August to the nineteenth of the same inclusively; and seen it, at the first like the Tail of a Scorpion; but on the nineteenth day resembling a Melon-seed."[11]

Our modern sunspot detective, John Eddy, extended this early historical fragmentary evidence by linking it to observations of the aurora borealis—the northern lights.[12] The northern lights are directly related to the level of solar activity and consequently to the number of sunspots. Eddy scientifically deduced that observations of the northern lights should independently match the records of sunspot number. Hence, numerous reports of northern lights should coincide with times of high sunspot numbers. Once again he turned to the historical records and discovered "between 1645 and 1708 not one [aurora] was reported in London."[13] Even in the northern latitudes where in modern times auroras are visible almost nightly, so few occurrences of the solar phenomenon were witnessed that they were considered portents.

Okay, so old records of sunspots numbers (dug up by Maunder and his wife) and of historical aurora observations (uncovered by Eddy) both indicate little solar activity. Remembering that scientists like to slam dunk their theories with abundant evidence, is there anything else supporting the idea of low solar activity between 1645 and 1715?

During a solar eclipse, one of the most memorable and impressive sights is the view of the solar corona—the sun's thin outer atmosphere. The shape of the solar corona varies according to the number of sunspots. The occurrence of many sunspots is related to the development of many pale—but very beautiful and noticeable—white coronal streamers extending outward from the sun's surface. If there are few sunspots, an observer sees only a faint, rather uniform *zodiacal light*, or false corona, around the black disk of the moon during the eclipse. Zodiacal light is not a part of the sun's atmosphere but simply the sunlight scattered by interplanetary dust.

So Eddy examined the records of sixty-three eclipses between 1645 and 1715 to see if they contained more descriptions of coronal streamers or of pale zodiacal light. Of course, many of those sixty-three eclipses occurred in remote areas where there were no seventeenth-century observers. But several did occur in Europe and were well documented. What did they show? According to Eddy, eclipse watchers at that time described the corona as "dull or mournful, and often as reddish,"[14] which was "consistent with our surmise of what the zodiacal light would look like at eclipse, were the true corona really gone."[15] He didn't find a single report of the whitish corona that is so well photographed during modern eclipses. Consequently, Eddy cited the uniformity of eclipse descriptions between 1645 and 1715 as his third pillar of evidence in confirming the lack of solar sunspot activity between those years.

For his fourth pillar of evidence, Eddy turned to one of our familiar climate sciences, dendrochronology—and, in particular, how tree rings relate to radiocarbon dating. Earlier I discussed the investigative tool of carbon-14 isotope dating—the idea of using radioactive carbon's natural and constant decay rate as a measure of determining age. One aspect of that dating method that I mentioned is a bit tricky is establishing how much ^{14}C exists at any given time. One reason for this potential problem is that the amount of ^{14}C depends on the quantity of cosmic rays—high-energy particles from the sun—impacting Earth's atmosphere, and the magnitude of cosmic rays is determined by solar activity. When the sun is highly active (lots of sunspots), its magnetic field shields Earth from some of the cosmic rays and less ^{14}C is formed, and the reverse is true under low solar activity: more ^{14}C is formed. Some of that ^{14}C is subsequently stored in the atmospheric carbon dioxide of that time—and eventually is locked into the cells of plants and animals.

Tree ring research can tell us how ^{14}C changes over time through a painstaking analysis of the specific amounts of ^{14}C in each tree ring (which is linked to the absorption of ^{14}C-rich carbon dioxide) compared to normal ^{12}C. So Eddy plotted the variations in ^{14}C abundance over the existing tree ring record (going back about five thousand years). What he discovered was actually more than he expected. He did indeed see a marked increase in ^{14}C in the rings of trees living during his target period between 1645 and 1715—this indicated a lack of solar activity for those years. That period has become known as the Maunder Minimum, in honor of the early solar detective E. W. Maunder's contributions to this interesting astronomical oddity. But Eddy also identified eight other periods of apparent low solar activity—a particularly strong one lasting from 1400 to 1510. That second low solar activity period is now known as the Spörer Minimum, after the second of those two early sunspot astronomers. An even earlier minimum, one lasting from 1280 to

1340, has been named the Wolf Minimum after Rudolf Wolf, who was so instrumental in early solar observations.

In summary, John Eddy demonstrated—convincingly through multiple lines of evidence—that the seventeenth (and the fifteenth centuries) were marked by low solar activity. But the question we now need to address is how did that absence of sunspots link to the climate of the time?

As I pointed out at the beginning of this mystery, the seventeenth century was bitterly cold and is the core of a time period now called the Little Ice Age. In fact, some of the coldest extremes experienced by civilized humans occurred in that time with global temperatures perhaps one and half degrees Celsius (three degrees Fahrenheit) colder than modern day. Could that frigidly cold period of the Little Ice Age—commonly coinciding with Western civilization's war and revolution—be the result of a prolonged absence of sunspots?

We must be very careful. Coincidences in climate are possible. Two things can happen at the same time and be totally unrelated. As Eddy himself noted, a colleague pointed out that the Maunder Minimum occurred during the reign of Louis XIV, le Roi Soleil—the Sun King.[16] Consequently, could one then conclude that the lack of sunspots created the rule of Louis XIV?

Remember one of the hallmarks of good science is multiple lines of evidence. The more that a scientist can establish a fact using independent sources of information, the more likely she can convince her colleagues (and the public) of her findings. The key point here is that Eddy proceeded to identify many different "solar activity minima" over the past five thousand years—the Maunder Solar Minimum was only the last of these.

Eddy next presented documentary evidence from climatologists that, in general, the climates of the other eight periods of solar activity minima over the last five thousand years were cold. But Eddy wrote his research in the mid-1970s. Today the case for "low sunspot number/cold Earth" isn't quite that straightforward. The problem? Earth's climate very seldom, if ever, responds as a single unit to any external change. For example, while Western Europe and eastern North America were undoubtedly colder than present-day during the Little Ice Age, climate reconstruction of the Central Asia region show that region was more moist (and rainier) than modern day.[17]

So today what scientists have undertaken is a calculation of the actual changes in solar radiation over the past compared to Earth's past climates. From that, they then entered those values into a general circulation model, which is a mathematical computer construction of all of the climate processes we understand on Earth.[18] What researchers found is that the small changes of solar radiation created by the Wolf, Spörer, and Maunder minima may have been enough to force down the average global temperature of the computer models to the point that they matched the changes in actual past temperatures. I should note, however, there have been some concerns voiced with the techniques used to re-create past solar receipt on Earth. Those concerns relate to how strong the temperature variability has been in the past (particularly in evaluating the warmth of a time called the Medieval Warm Period, which occurred just before the Little Ice Age).[19]

So our best evidence today suggests that Eddy—and through him, Maunder and Spörer—was most likely right, although, as with all of our mysteries, research continues. Fundamentally, we know that something odd had happened from the sixteenth to eighteenth centuries—at least with regard to Western Europe and eastern North America. The sun was very quiet and parts of Earth became very cold. A potentially troubling question arises: Could we see a new Maunder Minimum in the future? Unfortunately, that answer remains a mystery.

Time: January 1682
Location: A small monastery in western France, Europe

The Jesuit monk shivered in the intense chill of the morning and studied the bright round image being projected from the telescope onto a white wall behind him.

That is very unusual! There's now a second one!* The brown-robed monk thought to himself. *It appears as if the sun has taken a second bruise.

A second small black splotch was now apparent to the left of the first one that he had only just identified on the surface of the sun a few days ago. Both of the tiny splotches were about a third of the distance north between the equator and the north pole of the sun.

"Brother Dominic," he called out to his colleague. "Please look at this," he requested.

A tall monk with deep bushy eyebrows and a piercing gaze stepped over to the area on the roof observatory where the first stood next to the solar telescope.

"What is it, Brother Francisco?"

"A second spot has appeared on the sun. Have you ever heard of two spots appearing at the same time on the sun?"

Brother Dominic's bushy eyebrows rose slightly. "Two solar spots? Together? Simultaneously?" He gazed silently at the projected image of the sun on the white wall. "That is interesting, indeed."

"What does such an odd situation portend?" asked the first monk.

"By the grace of God, I do not know, Brother Francisco," answered the tall monk. "Our records of the sun at this observatory, which extend back for more than a century, do not ever show two sunspots appearing at the same time—indeed, the Most Excellent Signior Cassini at Paris says individual spots have only recently begun to reappear periodically on the sun after many years of absence—although some of the very ancient histories mention multitudes of such solar occurrences happening in far earlier times. I would suggest that we continue to carefully document the events. Perhaps someday, if we have enough accurate observations, we will ascertain their meaning."

Brother Francisco bowed deeply to the senior cleric and began to carefully write the description of the new sunspot into his records. A cold blast of wind whipped through the roof observatory, causing his fingers to shake slightly.

"How long must this bitter cold continue?" the monk muttered to himself as he carefully noted for the first time in perhaps two hundred years the occurrence of two sunspots at the same time.

Chapter 13

THE MYSTERY OF THE COMPLACENT EMPIRE'S WET WEATHER

As I mentioned earlier, sometimes the height of a civilization—its time of maturity—can be as interesting as its decline. An excellent case in point is the nearly three-hundred-year rule of the Qing dynasty in China.

As we have seen, the majority of European and western North American climate data, particularly the tree ring and historical records, indicated that the period of the Little Ice Age (1550–1850) was generally a time of significantly colder temperatures than those of modern day. Corresponding to this generally frigid environment, this period in Western civilization was, culturally, rather tumultuous, marked by colonial expansion, revolutions, and wars (for example, the colonization of the Americas, the American and French revolutions, and the rise of Napoleon).

In contrast, China and Southeast Asia at this time somewhat fell off the map. Over that time, from 1550 to 1850, only a few notable rebellions transpired in that region. The mystery for a climatologist is

Did weather aid in stabilizing Chinese society during the Little Ice Age? Was the lack of global attention due in part to a wetter set of climatic conditions than experienced by Europeans? Did a different climate in Asia during the Little Ice Age—compared to North America and Europe—play a role in the geopolitical stability of this region?

Time: July 1801
Location: The Summer Palace near Beijing, China, Asia
(This vignette is loosely based on one of the Memos to the emperor, translated by Q.-S. Ge and colleagues.)[1]

Emperor Jiaqing adjusted the silken sleeves of his gown and glanced down at the beautifully calligraphed parchment resting in his lap. He glanced over to his principal administrative aide, the elderly and wise councilor Zhu Hui. Zhu was a brilliant man whom he had just recently placed in charge of collating and evaluating the daily imperial memos, a task that, thus far in Jiaqing's short reign, the administrator had handled with great ability.

"The essential meaning of this?" the emperor asked, tapping the summary sheet.

"As my emperor has seen, I have independently confirmed the substance of the standard daily imperial memos," the old advisor answered carefully. "I have personally compared the values submitted by my emperor's officials from the surrounding regions. Given this summer's heavy rains, I have also dispatched several special investigators to parts of Zhili Province to verify those rainfall totals. The rains according to Jiang Sheng . . ."

"He is governor general for the Zhili Province, correct?" the emperor interrupted.

"Indeed, revered emperor. He reports that the past flooding rains, even by the end of last month, had deeply saturated the

farmlands in Zhengding. In a message I just received yesterday, my emperor's special agent at Shengzhu reports that recently a major thunderstorm broke over the area, lasting for an entire week. As my emperor knows, people in the Beijing area are experiencing major hardships due to this unrelenting flooding . . ."

The deep voice of Li Tzung interrupted, "The peasants can survive such hardship. They have before and they can again." The rather heartless comment came from the emperor's esteemed finance councilor—a relatively young but stout man of perhaps forty-five. Li Tzung shook his head in dismay at the aged imperial bureaucrat. "Heavy rain is an increasingly common occurrence throughout the empire. We must adapt to these conditions without unseemly obligation."

The emperor tapped his long, manicured fingernails on the reports carefully laid on his lap.

"What is the consensus from the imperial memos on the status of the agricultural grains?"

Zhu Hui considered the situation for a moment as he shuffled through his papers; finally the imperial bureaucrat raised his head at the emperor's stern face. "The reports are consistent, my emperor. They indicate that the massive flooding in Zhili Province will decimate this year's harvest for that region. The people around Beijing in particular are already suffering from a lack of food."

"And our options?" the emperor asked.

Zhu Hui took a deep breath. "I would humbly suggest the issuance of an imperial edict, my emperor. We need to release emergency funding for all of the people of the province or else there will be widespread starvation throughout the province by the fall season."

"I would strongly advice against such rash, impulsive action, revered emperor," countered the younger imperial accountant, Li Tzung. "The imperial treasury simply cannot support such a massive release of funds at this time."

Zhu Hui shook his head. "But if there is widespread

starvation—particularly here in Beijing, my emperor—the political situation could become troublesome. It could lead to even greater expenditures from the imperial coffers—remember the terrible flood of twenty-five years ago during your beloved father's reign?" the wise administrator responded.

The young imperial accountant hesitated—he had not considered that parallel to the present situation.

The emperor weighed the options in his mind. Although Jiaqing had not been emperor long—his father, Emperor Qianlong, had died only a couple of years ago—he had an intuitive feeling for politics and, perhaps more important, for the care and governance of his people. His two trusted administrators stood silently before him awaiting his decision.

Finally, the forty-year-old emperor spoke. "We will compromise. We will issue an imperial edict to release emergency funding without delay to relieve the people near Beijing. We must at all costs preserve order within our immediate area."

Both of his aides nodded in acknowledgment of the imperial voice of command.

"But," the emperor added, "as to how far to take such actions in the other parts of Zhili Province, we would like to have reports from our administrators for those regions to us as soon as possible. Once we have those reports and ascertain their needs, we will consider a further release of funds as required due to this emergency."

The two imperial administrators bowed low before him in obedience.

The Qing dynasty dominated China during the years 1644 to 1911 and marked the second time in history when the whole of China was ruled by foreigners. The first foreign rule was the Yuan dynasty (1271–1368) when the Mongols, starting with Genghis Khan, conquered and then controlled the region. The Qing dynasty, in contrast to that rather tremulous rule, governed with little opposition or oppression throughout China. In particular, the reigns of the first three emperors of the Qing dynasty were characterized by prosperity and relative peace throughout the land. A consensus of many historians appears to be that these first three rulers provided strong and consistent leadership for more than 130 years.

During this time, the Qing emperors created demonstrable growth in many areas of society. For example, roads and other public works were repaired and maintained. Yet, overall, taxes were kept very light, at least compared to previous imperial periods in China's history. This was due in part to the growth of commerce and international trade. Indeed, the Qing emperors actively rewarded greater land cultivation by reducing taxes or exempting those lands from taxation. Such policies greatly promoted economic growth, particularly at the frontiers of the country. The social stability led to a revival of the arts and learning. The emperors commissioned literary works on many aspects of society. Artistic endeavors during the Qing dynasty included the development of new colors in various types of porcelain. In the realm of literature, the reigns of emperors Kangxi and Qianlong saw the compilation of several major works such as the *Encyclopedia of Chinese Writings* (containing Confucian classics, history, and philosophy), the *Kangxi Dictionary*, and *A Collection of Books Ancient and Modern*. However, it should be noted that the Qing rulers also banned and destroyed many works that did not receive their explicit approval.

In general, the Qing emperors looked inward more than outward. Their foreign policy was one marked by conservatism and isolationism. Although European missionaries were allowed into China during the early years of the Qing dynasty, they were later banned. This was due in part to accusations of Christian sailors looting the Chinese coast and growing disputes among the missionaries. And, throughout the early times of the Qing dynasty, the borders of the empire grew. By the late 1700s, Taiwan became part of the country and the border with Russia was finally fixed. The Qing emperors then concentrated on resolving long-standing internal conflicts such as those between the nomadic and peasant populations. They consequently instituted measures to develop the economy, culture, and transportation infrastructure in the frontier areas. As a result of these changes, they laid the foundation for modern China's territorial boundaries. And, with this new prosperity, the Qing power structure became more centralized, allowing for an improvement in the social order and the growth of the population to perhaps 300 million by the start of the nineteenth century.

With such a prosperous and civilized—yet spatially extensive—society, it would seem that documentation of the weather during the Qing dynasty would have been possible—and indeed the Qing rulers did order and receive such reports. But, interestingly, it has only been recently that climate scientists have uncovered that treasure trove of climatic information and begun to use it in climate research. The Mystery of the Complacent Empire's Wet Weather can be uncovered in part by the study of the Qing emperors' official weather reports.

A large empire—in both geographic size and population—must by necessity employ a hefty bureaucracy. Bureaucracies in turn create enormous mountains of paperwork. The Qing dynasty was no exception to this maxim. During the time of the Qing emperors, "Memos to the Emperor," which reported daily administration activities, were demanded of the provincial and local government officials around the country. Indeed, during the Qing dynasty, a special office was maintained outside the Qianging Gate of the Forbidden City just to receive the "Memos to the Emperor" each and every morning. These memos were hand delivered by special couriers from the provinces and then conveyed directly to the emperor. Many of these reports contained specific accounts of weather—in some cases, very detailed weather information.[2] For example, in 1736, Emperor Qianlong's first year in power, local administrators were required to document daily information of *Yu-Xue-Fen-Cun*. Literally, that is *Yu* (rainfall)—*Xue* (snowfall)—*Fen* (a unit of measurement, approximately an eighth of an inch)—*Cun* (Ten Fens). This meteorological reporting continued on a relatively constant basis throughout the Qing dynasty for the next nearly three hundred years, until 1911.

The geographic area covered by these reports was extensive. A total of 273 administrative sites within eighteen provinces and the special region of Shengjing (in Mongolia, north of China) were part of the administrative memo network. The quality control imposed by the emperor was equally impressive. Each of the Yu-Xue-Fen-Cun records had to be submitted concurrently—but independently—by administrative, educational, and military officials so that the emperor could crosscheck and verify the accuracy of the records. For example, in one

"Memo to the Emperor," Aertai the governor-general of Shandong Province, reported, "Every country in Taian Fu received rain infiltrating into the arable layer [a concept written as "Shen Tou" in Chinese] on the 23rd [of] the fourth month in lunar calendar of the twenty-seventh year of Qianlong Emperator [May 16, 1762]; other Fu's received the rainfall around the end of the fourth month, and the soil is very wet."[3] The emperor curtly responded to this seemingly precise and detailed report with the following comment: "Your description is suspicious. The rain infiltration depth is less than 1 cun in Tain reported by Wang Jiongu. Then why was it described in your report as Shen Tou?"[4] In other words, the emperor noted that there was a marked discrepancy between two separate weather reports for the same region. Somebody was likely lying.

Such imperial scrutiny inspired the local administrators to undertake direct investigations of the weather on their own for fear of imperial disfavor if they failed: "Reported by Li Kentang, Governor General of Zhili province . . . there was little rain after the end of the third month. . . . On the 25th the fourth month . . . it was raining heavily. I went to farmland to dig into the ground by myself, and the depth of infiltration has reached 4 cuns."[5] In addition, the emperor even occasionally employed special high-ranking officials—in essence, weather spies—to conduct special investigations.

One difficulty with the interpretation of such detailed records is reconciling the unit of measurement to today's values. The critical information that the emperors required from the regional officials was the *Yu-Fen-Cun*—the soil penetration depth—a measure that indicated how deeply the water had soaked into the ground. Although by this time in the West rain gauges had been invented, rain gauge measurements were not given during the Qing dynasty. The emperors were more concerned with how weather was affecting agriculture rather than the actual rain amounts. That creates a modern-day dilemma. In order to use and compare the detailed "Memos to the Emperor" records, the *Yu-Fen-Cun* values need to be converted into contemporary rainfall estimates. Luckily, our modern climatologists proved themselves up to the task. Researchers from China and the State University of New York at Albany devised an ingenious modern field experiment to ascertain exactly how much rain water would have to fall in order to correspond to the given measurements of Yu-Fen-Cun, or soil penetration.[6]

The experiment was performed at an agriculture experimental station in Shijiazhuang, China. The researchers instrumented a wheat field with an artificial rainmaking machine. They then conducted a series of experiments, accounting for rain intensity and amount as well as the soil conditions (ranging from very dry to wet), in which they determined the exact soil penetration values for specific rainfall amounts. They statistically analyzed data from a total of forty-one tests and created a regression equation linking the rainfall to the depth of penetration of water into the soil (Yu-Fen-Cun). This statistical equation accounted for 87 percent of the variability of the test data—meaning it did a markedly good job of accurately predicting how much rain would produce a given amount of soil penetration. In other words, if we know the soil penetration amount, we can determine the rainfall that produced it. So, with the experimental equation, the researchers went back to the actual Yu-Fen-Cun records and reconstructed the rainfall (and snowfall) amounts for a large part of eastern Asia from 1736 to 1911. And what did that time series of precipitation show?

Remember that a large portion of the Qing dynasty corresponded to the period termed the Little Ice Age. In that period, as discussed in the last mystery, most of the earth—but in particular, eastern North America and Europe—experienced much colder and harsher conditions than exist today.

In China, however, the key difference wasn't temperature so much as it was rainfall. The reconstructed rainfall record—based in part on the "Memos to the Emperor"—indicates that, in contrast to the harsh conditions of Europe during the Little Ice Age period, both summer and winter precipitation rates in China were significantly higher than present-day averages.[7] The ample rainfalls—particularly winter rains—in the Little Ice Age for China likely led to consistently good harvests, and perhaps promoted a more prosperous and contented society.

This extra rainfall also created a great propensity for floods for at least a part of China. A flood reconstruction study by Chun Chang Huang and five colleagues from China's Shaanxi University based on detailed analyses of riverbank sediments indicated that a majority of the floods over the past 2,200 years on the Sushui River in the Yuncheng Basin occurred in the Little Ice Age between the fourteenth and nineteenth centuries. They identified particularly large floods for the years 1570, 1662, 1745, and 1761.[8]

Only by the end of the Little Ice Age around 1850 or so did we begin to see a consistent marked decline in winter precipitation over China (although I should note that a few serious droughts nevertheless did occur throughout the time of the Qing dynasty, such as the drought in 1792).[9] That decline in winter rains was coupled with a decrease in summer precipitation by 1890. Sudden, periodically short-term increases in rainfall occasionally broke the increasing aridity. Such dramatic environmental shifts became horribly noteworthy in the post–Little Ice Age histories of China.[10] A drought that some historians have labeled as the worst natural disaster in the recorded history of the entire world occurred in 1877.

Because of the massive population that had developed in this fertile area during the Qing dynasty, any sudden drop in rainfall could prove catastrophic. The drought of 1877 resulted in a lethal famine that killed between 9.5 and 13 million people. According to a report from the Baptist missionary Timothy Richard, people began to eat "elm bark, buckwheat stalks, turnip leaves, and grass seeds. When these are exhausted, they pull down their houses, sell their timber, and it is reported everywhere that many eat the rotten kaoling reeds [sorgum stalks] from the roof. . . . Thousands eat them, and thousands die because they cannot get even that. They sell their clothes and children."[11]

In one area during this terrible drought, the only means of disposing of the dead was to bury them in huge pits, which are still today called "ten-thousand-men holes." Although some accounts of cannibalism had been reported in other famine-hit provinces, the practice of eating human flesh became especially widespread in Shansi. The *New York Times* of July 6, 1878, quoted a letter from the Roman Catholic bishop of Shansi, Monsignor Louis Monagatta, in which he reported, "Until lately the starving people were content to feed on the dead; but now they are slaughtering the living for food. The husband eats his wife; parents are eating their children; and in their turn sons and daughters eat their dead parents. This goes on almost every day."[12]

The imperial bureaucracy was simply unable to handle a disaster of that magnitude. Relief supplies destined for hard-struck Shansi Province were met with major logistical problems. While the city of Hwailuhiern, the starting point for relief supplies, was filled with officials who were intent on getting relief convoys over the pass, the area also swarmed with fugitives, beggars, and thieves intent on looting the convoys. Along the convoy routes, so many camels, oxen, mules, and donkeys were pushed beyond their limits that many perished or were even killed by the desperate people in the hills for food. Reports described the road as being marked by the rotting carcasses of both men and beasts.

The governmental collapse continued even when the weather conditions reversed themselves—and the floods came. Throughout the Qing dynasty, the abundant Chinese rains of the Little Ice Age climate did on occasion produce massive floods, which the governmental infrastructure was able to handle. When disasters began to occur more frequently, however, the government could not address them adequately. Such was the case when the Yellow River, swollen by abnormally heavy rains, broke through the dikes in the heavily populated province of Honan a mere decade after the horrendous drought of 1878. The waters tore a nearly three-hundred-foot gap in the dikes. That break quickly widened to over half a mile—and the flood waters poured into the central plains of China.[13]

At its height, the 1887 flood covered 480 kilometers (roughly 300 miles) and flooded 11 cities and 15,000 villages. The North China *Herald* reported that a dead child had floated to shore on the top of a chest, where it had been placed for safety by its parents with food and name attached. In another place, a family, all dead, were found with the youngest child placed on the highest spot on a couch. According to some, the waters killed nine hundred thousand people with an equal number succumbing to disease and starvation. Others place the death toll as high as perhaps six million people. These latter death estimates would rank the Great China Flood of 1887 as the worst such calamity in recorded world history.

A missionary from the area wrote this description of the disaster: "In Cho-chia-kow itself fifty streets were swept away, leaving only three business streets, on the north side, which are all flooded. The west and south parts of the city are on opposite sides of the stream. The whole area is one raging sea, ten to thirty feet deep, where there was, only a month ago, a densely populated rich plain. The newly gathered crops, houses and trees, area all swept away, involving a fearful loss of life and complete destruction of next year's harvest. The river is all coming this way now, and a racing, mad river it is. The mass of the people is still being increased by continual arrivals, even more wretched than the last. There they sit, stunned, hungry, stupid, and dejected, without a rag to wear or a morsel of food."[14]

As China and its Qing leaders were struggling in the post–Little Ice Age world with a series of environmental disasters, they also had to deal with the growing influence of suddenly flourishing Western countries. The European countries had survived the Little Ice Age and witnessed the growth of colonization and industrialization. The relatively harsh conditions in North America and Europe between 1550 and 1850 helped—in part—to promote the growth of mechanization partially as a means for overcoming the environment. The Western-based Industrial Revolution did not, for the most part, reach China. Instead, in the more drought/famine ravaged times of the mid- to late-nineteenth century following the Little Ice Age, the Qing government was forced to sign a series of markedly inequitable treaties with the European powers, which mandated that China had to cede territories, pay indemnities, and open trading ports. Eventually China became, in essence, a semifeudal country. The central Qing government (with younger and younger emperors lacking direct control) was forced to impose more taxes in order to pay both the expenses of war and the indemnities they had to bear. Eventually, this sparked a series of antifeudal and anti-imperialist movements such as the Taiping Rebellion—which led to growing unrest in the population. Although the Qing government attempted to introduce some reforms, in 1911 a revolution led by Dr. Sun Yat-sen enabled the Chinese people to overthrow the Qing emperor.

The Qing dynasty in China had prospered during the relatively moist environment of the Little Ice Age, but the different environmental conditions that followed that wet period in China's history had, in part, perhaps proven too great for the empire to survive. And that variable environmental trend continued into the twentieth century.

Time/Location: 1911, China. Flooding from the Yangtze River has led to the deaths of more than a hundred thousand from drowning, another hundred thousand from starvation, and many more murdered by roving bands. Western missionaries in the country report the bizarre sight of thousands of wooden coffins floating down the river following the inundation of the largest cemeteries . . .[15]

Time/Location: 1915, China. Flooding of the Kwangtung and Kiangsi rivers kills over a hundred thousand people . . .[16]

Time/Location: 1922, China. A typhoon, labeled as one of the worst in history in the China Seas, crosses the Chinese coast with torrential rains and incredible winds. The greatest damage is done by an enormous sea wave, which washes away all the houses that were not been blown down. Out of a population of sixty-five thousand persons, it is estimated that fifty thousand lost their lives. It will be several days before the water is drained off the countryside . . .[17]

Time/Location: 1928, China. A massive drought has struck the three central Chinese provinces with an estimated death toll of three million . . .[18]

Time/Location: 1931, China. Floods of the Hwang Ho River are reputed to be responsible for over three and half million deaths . . .[19]

Time/Location: 1936, China. Famine brought on by a terrible drought has left five million dead . . .[20]

Time/Location: 1938, China. The Yellow River flooding has killed one million in 1938 during the Japanese Army's advance into China. Chiang Kai-shek orders levees on the Huang He River to be dynamited . . .[21]

Time/Location: 1939, China. A persistent drought over the central provinces has left over two hundred thousand dead and another twenty-five million left homeless . . .[22]

Chapter 14

THE MYSTERY OF THE PACIFIC HOT TUB

Mysteries bring out the detective in all of us. Discover something that appears baffling, and the likelihood is that fairly quickly someone will attempt to explain it. There is simply an innate curiosity in humans to supply answers to mysteries. So when someone sees something that appears to go beyond the obvious, often many people will quickly investigate it. However, that was not the case with the mystery first put forth by the illustrious naturalist and author of the landmark evolutionary book *On the Origin of Species*, Charles Darwin. While Charles Darwin first posed the following climate mystery in the 1830s, its eventual solution took nearly a century and a half.

Time: December 29, 1833
Location: Onboard the HMS *Beagle* near the coast of Argentina, South America
(This and the closing vignette of this chapter are

loosely based on entries recorded in the journal of Charles Darwin.)

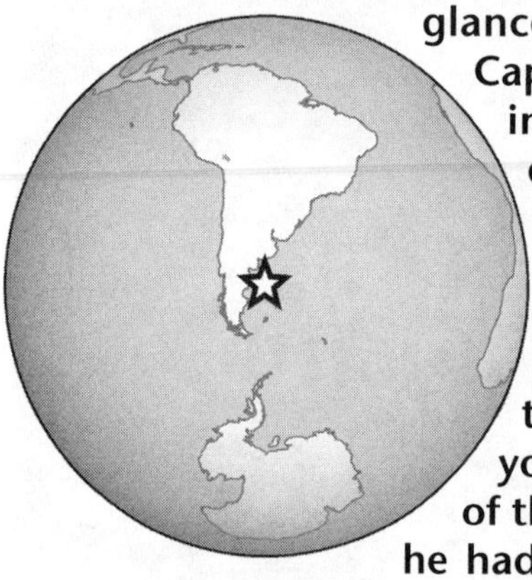

The young naturalist coughed slightly as he puffed on his pipe a bit nervously. With a shiver of trepidation, he glanced anxiously at the legendary Captain Robert FitzRoy. The two sat in the captain's relatively small quarters—the *Beagle* was not a large ship—discussing the recent events and discoveries of the voyage. The naturalist fought to stem his nervousness. The captain was—somewhat to the youthful scientist's surprise—one of the most knowledgeable men that he had ever met. He was beginning to realize that these evening talks were key reasons why Captain FitzRoy had selected him for this monumental voyage around the world. After all, he was only a very young scientist with distressingly few academic credentials. Apparently, for this voyage, the captain expected this fresh young scientist to be ever observant and innovative—and to be able to relay those observations to the master of the ship in an intelligent manner.

"I have found it to be particularly fascinating, sir," the naturalist began tentatively, "about this so-called *gran seco*, of which I have heard repeatedly while in Argentina."

The captain, also smoking a long pipe, looked up from his logbook with a slight flicker of interest. "Gran seco? Pray tell, Mr. Darwin, what is that?"

"Apparently, sir," the naturalist elaborated with an inward sigh of relief at the success of his conversational gambit, "a gran seco refers to a drought of epic proportions. According to the locals, so little rain fell between the years 1827 and 1830 that the vegetation, even the thistles, failed; the brooks were dried up, and the whole country assumed the appearance of a dusty high road."

"Indeed?" the captain asked as he closed his logbook to devote full attention to the conversation. The naturalist had discovered early during this world voyage that the captain loved to talk about the weather.

"Yes," the young naturalist nodded his head in growing enthusiasm. "Particularly in the northern part of the province of Buenos Ayres, a very great number of birds, wild animals, cattle, and horses perished for the want of food and water. One local told me that the lowest estimation of the loss of cattle in that province alone was taken at one million head."

"A million head of cattle?" the captain asked, disbelievingly.

The young scientist nodded eagerly. "I was informed by a reliable eyewitness that the cattle in herds of thousands rushed into the Parana—that is one of Argentina's main rivers."

Captain FitzRoy motioned for the young man to continue.

"Well, these cattle were so exhausted by hunger from the drought that they were unable to scramble up the muddy banks, and thus were drowned. The same thing is said to have happened in fairly regular intervals of about every fifteen years or so in the past. Very odd to think of cattle drowning during a drought."

FitzRoy rubbed his chin thoughtfully. "Even more interesting is the possibility of an inherent periodicity in these droughts. I have noticed that periodical droughts are somewhat more common in dry climates than in wet—such certainly is the case for Australia. I recall that Captain Sturt wrote that droughts recur there every ten to twelve years, and then are followed by many years of excessive rains till the next drought."

The young naturalist quickly agreed. "Yes, Captain. I have read that the year 1826 and the two years following were singularly dry in Australia." He paused a moment. "That would coincide with the first of the gran seco droughts here in Argentina."

FitzRoy closed his eyes and relit his pipe. The room was filling with aromatic smoke—a good sign, the young scientist realized with growing delight, of an engaging conversation for the captain. Such conversations, the naturalist knew, helped the captain combat his foul fits of temper.

"Hmm," the captain muttered thoughtfully. "Have there been other coincidences of simultaneous drought around the world? I seem to recall something around the late 1700s . . ." His voice trailed off as he stopped to puff on his pipe. The captain had a slight gleam in his eye as he looked at his young apprentice scientist.

"Yes, you are certainly right, sir," the brash, young naturalist grinned as he seized the opportunity to impress his captain. "General Beatson, in his account on the natural environment of the island of St. Helena, wrote that he witnessed a severe drought on that island around 1791 but that the aridity of that year was apparently far more calamitous in the more distant land of India."

The captain looked up thoughtfully. "And, if I am not mistaken, did not Mr. Bryan Edward—in his *History of the West Indies*, as I recall—write that the winter season between 1791 and 1792 was unusually dry on the island of Montserrat in the Caribbean Sea—half a world away from India?"

The naturalist nodded eagerly, totally caught up in the growing string of coincidences. "When I was reviewing the literature on the Verde Islands last year, I remember reading that Mr. Barrow in the latter part of 1792 wrote that a drought of three years' continuance—and consequent famine—had nearly desolated the main island there."

"Fascinating," the captain said.

A gleam appeared in the eyes of the naturalist.

"I wonder," the brilliant youthful scientist speculated, "if there might be some very general cause that

is creating these widely separated droughts around the world to occur simultaneously?"

Charles Darwin put forth a very interesting mystery: Is there a very general cause to many of the world's droughts?[1] It has been only recently that modern climatologists have solved that mystery. We discovered that there is indeed one very general cause for all of those droughts that Darwin mentioned—and many other floods and droughts that have followed. We now call that cause the *El Niño–Southern Oscillation*, or ENSO, phenomenon.

The route of the HMS *Beagle* and Charles Darwin in the early 1830s. Map by Barbara Trapido-Lurie.

So our first major question involving this mystery is simply What is ENSO? The term incorporates the two distinct aspects of the phenomenon. The first is oceanic: EN stands for El Niño, an abnormally warm current (and its cold current counterpart, La Niña) in the eastern Pacific Ocean. The other aspect of ENSO is atmospheric: SO represents the Southern Oscillation, a wind pattern over the southern Pacific Ocean. The more famous (or infamous) of ENSO's two aspects, El Niño, is derived from a Spanish term meaning "the Child"—specifically, the Christ Child, referring to the time of the year when El Niño becomes most noticeable, principally around Christmas in the months of December and January. El Niño historically refers to a marked warming of the coastal waters off of Peru and Ecuador in South America. The phenomenon is evidenced in that region by torrential rainfall, often resulting in catastrophic flooding, and a ruinous collapse of the anchovy fishing industry, an economic mainstay of the region. El Niño events in South America have been well documented back to 1726 and there is other evidence indicating occurrences for at least a thousand years prior to that. Paleoclimatic research has even suggested the possibility that El Niño events may have occurred as far back as five thousand or more years ago.[2]

In contrast, the atmospheric aspect of ENSO, the Southern Oscillation, refers to the subtropical airflow that exists over the southern Pacific Ocean, specifically between the coasts of western South America and Australia. The Southern Oscillation experienced a more recent discovery than its oceanic counterpart. More than a half century after Charles Darwin in the early 1900s, Sir Gilbert Walker became one of the earliest investigators of the Southern Oscillation and its impact on climate around the globe.[3] Walker was one of the first scientists to recognize that an atmospheric circulation cell, that is, a looping circle of air that travels back to its starting point, exists in the southern Pacific Ocean between Australia and the west coast of South America.

Walker determined air that is forced by the high pressure located over the eastern Pacific normally travels westward across the cold ocean surface to the warmer waters of Australia and Indonesia. There the air rises into the upper atmosphere, flows back toward the east, and then sinks down to the surface over the eastern Pacific. This creates a closed loop of air, or a circulation cell. While the discovery of this loop of air was important to many aspects of meteorology, Walker failed to link it to the oceanic element of the phenomenon, El Niño.

It was only in the 1960s that significant scientific work connecting the oceanic and atmospheric elements of ENSO was undertaken. That work was spearheaded by a renowned meteorologist named Jacob Bjerknes (the son of one of the creators of polar front theory discussed earlier in the Mysteries of Weather and Climate in Ancient Greece). Bjerknes addressed three main attributes of the oceanic circulation in the Pacific Ocean.

- *Upwelling*: The rising of water toward the surface in a body of ocean. In general, upwelling is most prominent when winds blow parallel to a coast. Upwelled water is both significantly colder and richer in nutrients than the surface water. The cold temperatures and the corresponding high-nutrient content (critical to the anchovy industry) of the waters off the west coast of South America are linked to upwelling.
- *Thermocline*: A vertical temperature gradient in a body of water that is markedly larger than the temperature differences directly above and below it. Often the thermocline serves as a barrier to inhibit movement through it.
- *Convection* (or *adiabatic uplift*): The uplift of air due to surface heating. This process often creates thunderstorms and rain.[4]

Bjerknes discovered that the sea-surface temperatures at the eastern end of the Pacific Ocean (near South America) are much colder than one would expect given the subtropical latitude of the area. Conversely, the western Pacific Ocean waters (near Australia and Indonesia) are very warm. Bjerknes reasoned that there must be a large east-west temperature gradient along the equator in the Pacific Ocean.

Since a key fact of nature is that warm air rises, Bjerknes reasoned that the air must rise above the warm air over the hot waters of Indonesia, flow eastward, and then sink down to the chilly waters off the west coast of South America. The relatively cold, dry air above the cold waters of the eastern equatorial Pacific must then flow westward along the surface toward the warm west Pacific. "There," Bjerknes wrote of the west Pacific, "after having been heated and supplied with moisture from the warm waters, the equatorial air can take part in [a] large-scale, moist adiabatic [convective] ascent."[5] Some of the ascending air, he conjectured, joins the poleward flow at upper levels associated with the Hadley (or north-south) circulation and some returns to the east to sink over the eastern Pacific. Since rising motion is associated with low pressure and sinking motion with high pressure, Bjerknes identified two distinct pressure systems over the South Pacific, a low-pressure system near Australia and a high-pressure system to the west of South America. He named the airflow between those two pressure systems—in honor of the previous El Niño researcher—the Walker Circulation.[6] So Bjerknes had linked the oceanic and atmospheric parts of ENSO.

This left a critical question still unanswered. What causes the temperature difference between the east and the west sides of the Pacific Ocean in the first place?

Bjerknes suggested three possible causes for the unusually cold sea-surface temperatures in the east Pacific:

- *Near-equatorial upwelling.* The rotation of Earth (the Coriolis effect) forces ocean currents to turn to the right in the Northern Hemisphere and to the left in the Southern Hemisphere. Consequently, the surface flow is deflected toward the equator along South America and that flow of water must be fed by waters that upwell along the coast, waters that are colder than the surface.
- *Horizontal advection.* The southerly winds drive currents that cause cold water to be pulled from Antarctica along the South American west coast, then eastward along the equatorial Pacific.
- *Upward thermocline displacement.* The tropical ocean has two distinct layers: a shallow layer of warm waters resting above a layer of deep cold waters. These two layers are separated by the thermocline, a narrow (roughly half a football field deep) region marked by a strong temperature change (18°F or more). The surface easterly winds along the equator push the waters of the warm upper layer to the west and then poleward, pulling the thermocline to the surface in the east. As a result, the water upwelled there is colder than it would

> be if the upper layer waters were more evenly distributed across the Pacific Ocean.[7]

Bjerknes then asked which of these three factors is the predominant cause of the temperature difference across the Pacific. In particular, which one is most important in creating a colder eastern Pacific Ocean?

Unfortunately, even today, we don't know for sure—although it seems to be the case that all three have some effect on ocean temperatures. However, even though he didn't answer that fundamental question, what Bjerknes did accomplish was impressive: he determined that the oceanic and atmospheric circulations were mutually supportive in a chain reaction. He described the chain reaction as "an intensifying Walker Circulation [that] also provides for an increase of the east-west temperature contrast that is the cause of the Walker Circulation in the first place."[8] This sequence of events creates the occurrence of cold equatorial waters in the Pacific that we now call La Niña—the opposite of El Niño.

Bjerknes also noted that the ocean/atmosphere interaction could operate in the contrary sense; a decrease of the equatorial easterly surface winds diminishes the supply of the cold waters to the eastern equatorial Pacific (by any of three mechanisms). The lessened east-west temperature contrast causes the Walker Circulation to slow down—and creates El Niño, a warm flow of equatorial water toward South America.

So Bjerknes discovered that ENSO is fundamentally a huge positive-feedback phenomenon—the ocean/atmosphere components work to amplify each other. He also determined that there are two distinct phases: El Niño (when the two work together to create warm eastern Pacific waters) and La Niña (where they combine to produce cold eastern Pacific waters).

Unfortunately, Bjerknes didn't address the next critical issue needed to understand and forecast El Niño: What are the mechanisms that control equatorial water/air movement in one phase (El Niño or La Niña) or the other?

Luckily, a professor at the University of Hawaii, Klaus Wyrtki, was able to provide the next piece to the puzzle. He knew that El Niño, the equatorial warming of surface waters in the Pacific, occurred primarily due to air flow rather than heating changes. So instead of examining the ocean water temperatures, he analyzed sea level changes (suggesting that the winds "push" the waters back and forth).[9] But how can one determine sea level changes around the world? After all, the long-term changes in sea level in some places around the world are often less than an inch—how can one measure such small variations across all of the world's oceans?

Part of the solution rested with a pair of unique satellites, the first called TOPEX/Poseidon, launched in 1992, and its successor named Jason-1.[10] These satellites employed incredibly sensitive altimeters on board to measure the surface of the global ocean based on the height of the satellite. The accuracy of these instruments allowed the average sea level around the world to be estimated to an astounding precision of several millimeters (less than an eighth of an inch). Consequently, these altimeters provided levels of precision previously unobtainable in global sea level measurements. These satellites were able to detect small movements of water in the sea on spatial scales of twenty miles or more. Such changes in ocean topography were precise enough to detect a mere four-inch change over six hundred miles during an El Niño event in the tropical Pacific.

Using sea level data from such satellites as well as other data from across the Pacific Basin, Wyrtki discovered that, although sea-surface temperature changes were primarily limited to the eastern basin, the actual sea level changes were basin-wide; in other words, they spanned the whole ocean from South America to Indonesia. And, moreover, he discovered that the places in which the initial wind changes were causing sea level changes were in the central and western Pacific Ocean. This is opposite from the locale of the major temperature changes, which occur primarily in the eastern Pacific. Wyrtki therefore suggested the cause of the eastward movement of the ENSO signal from the region of Australia to the west coast of South America was not because of temperature gradient changes but through the medium of equatorial Kelvin waves—actual movements of water across the ocean.

The concept of an oceanic Kelvin wave is similar to that of an atmospheric wave (the kind of path that the atmospheric jet stream makes, commonly called a *Rossby wave*). However, oceanic Kelvin waves are limited to the equatorial regions and operate in response in part to energy/mass transfers by atmospheric waves. In contrast, Rossby waves involve energy transfer resulting from Earth's rotation

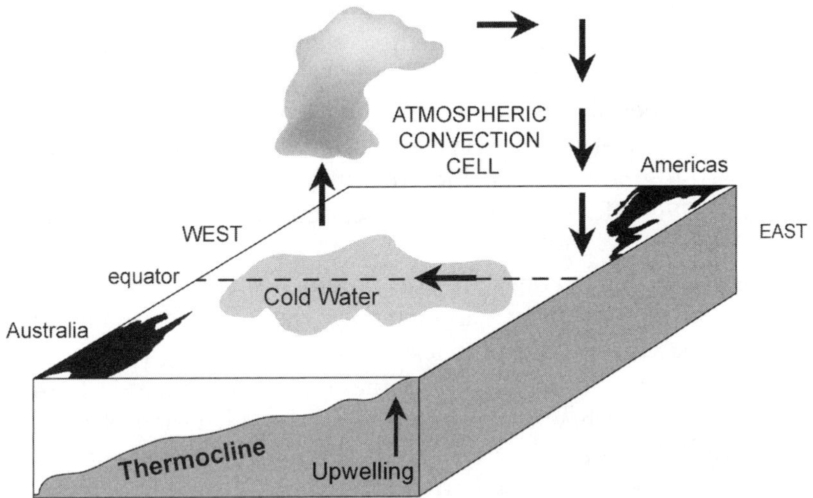

The important physical characteristics of the El Niño–Southern Oscillation, or ENSO, phenomenon. This figure shows the more common La Niña situation. Graphic by Becky Eden.

(the Coriolis effect). One critical difference with regard to ENSO is that Rossby waves move energy westward (toward Australia) while Kelvin waves propagate eastward (toward South America). Another important difference is that Rossby waves move much slower (by a factor of three) than Kelvin waves. The Rossby waves transfer water and energy westward (La Niña) until they hit a boundary—like the Australian and Indonesian coastlines—at which time the water is eventually returned eastward in the form of faster-moving Kelvin waves (El Niño). These waves can be expressed as mathematical equations.

Given these findings, scientists were able to create computer models displaying the main aspects of ENSO. Those early ENSO models indicated that a recharging of the equatorial reservoir of warm water is critical for the creation of El Niño. Why? The aftermath of El Niño leaves the thermocline along the equator shallower than normal—in essence, only a small amount of warm water exists in the western Pacific after an El Niño event pushes the hot water of that area eastward to South America. It therefore takes the next few years to gradually refill the equatorial warm water reservoir to the point that a Kelvin wave can once again transport the warm water eastward.

Only when there is enough warm water in the western Pacific near Australia and Indonesia can the Kelvin waves move enough warm water to the eastern end of the equatorial Pacific. This explains why generally El Niño is a periodic event of normally about seven years or so—since a very distinct recharge time is needed. Recent past El Niño events include the ones in 1976–77, 1982–83, 1987–88, 1990–94 (the longest extended period on record), 1997–98, and 2006–2007.

Therefore El Niño–Southern Oscillation might be explained by using a hot tub analogy. Waves and winds on the surface can cause the water of the hot tub to "slosh" back and forth, changing the level of the water from one side of the tub to the other. With one side of the hot tub holding much warmer waters than the other side, a major temperature shift in ocean waters occurs when the slosh happens.

But, if Charles Darwin was right that there is some "very general cause" that connects the weather around the world and this hot tub slosh effect is responsible, how does it work? This critical question involves ENSO's role in creating climatic change around the globe. How can a watery "slosh" of the Pacific Ocean create distant changes in weather around the entire world?

Because the Pacific Ocean is so large, the airflow over the ocean is critical to the basic winds around the entire world. When the Pacific airflow is disrupted by El Niño, the entire global wind circulation changes, including the location of such things as the jet stream. That in turn causes variations in atmospheric features such as the location of the storm track—and those changes can create flooding rains in places that might normally not get them or pronounced drought in places where the rains normally fall. The key is that atmospheric winds around the world are all linked together—change the winds in a big enough part of the world and the whole global atmosphere will adjust to the change.

For climatologists, El Niño–Southern Oscillation is a *climatic teleconnection*.[11] A climatic teleconnection involves a set of weather and/or oceanic phenomena that occur in one part of the world that subsequently affects a distant portion of the world. For example, the consequences of the 1997–98 El Niño event on the weather were profoundly global.[12] Some of the dramatic regional effects associated with El Niño included massive floods in Peru, drought and wildfires in Australia, drought in Brazil, very heavy rains and flooding throughout the southern United States, fewer Atlantic hurricanes, and a failure of the Southeast Asian Monsoon. Knowledge of the Pacific Ocean's weather is therefore critical to the rest of the world.

So we have gained a pretty good conceptual understanding of how ENSO works. However, some unanswered questions still remain. Primary among of them is what actually initiates the abrupt transition from the westward movement of cold La Niña waters to the eastward "slosh" of a warm El Niño event? At this time, we don't have the answer—everything from undersea volcanoes to lunar tidal forces has been hypothesized to initiate the shifts.[13] One thing we know, however, is that once the shift starts to occur we can accurately forecast its effects around the entire world. We have learned—as Charles Darwin had speculated so long ago—that there is indeed "a very general cause" for much of the world's weather.

Time: September 14, 1835
Location: Several hundred miles off the coast of western South America aboard the HMS *Beagle*

The naturalist sat in his tiny cabin, already tightly crammed with a multitude of biological and geologic samples from the places that he had visited during the long voyage of the *Beagle*. Still somewhat uneasy with the gentle swaying of the ship, he hunched over his tiny desk, idly flipping through the pages of his diary. It was nice to have the time to recapture some of his thoughts and ideas over the last two years. The *Beagle* at this time was still between ports, far out to sea, and it was a good time to catch up on his notes. Particularly since Captain FitzRoy was in one of his foul angry moods, he knew not to gain the master sailor's attention.

The young scientist turned his diary's pages back to the evening conversation he had enjoyed with the captain several months ago regarding the gran seco—the great drought—that Argentina had recently suffered. Ah, yes, so much had happened since then that the young man had almost forgotten that fascinating discussion of simultaneous droughts around the world.

What had he written? Ah, yes, there might be

"some general cause" for all of those droughts around the world. That was an intriguing idea. He was glad that he had jotted down a note about it.

"An exceptionally interesting hypothesis," he muttered softly to himself. "But, if there is indeed some general cause of these disparate droughts around the world, what could it be? I would imagine that the culprit would likely be either the currents of the air or the currents of the sea—those two aspects of nature are universal across the globe."

He pondered the two possibilities for a moment.

Air currents seemed plausible mechanisms for creating drought—after all, hadn't the American scientist Franklin hypothesized a few decades ago that weather traveled from one place to another in response to the winds? But the young naturalist smiled as he played devil's advocate to his initial thought. It was equally likely that changing ocean currents could be the possible agent of simultaneous drought—the temperatures of his ocean water samples here off the west coast of South America were significantly colder than the ocean waters that he had collected along the east coast. Perhaps variations in ocean water temperatures might produce drought.

What a fascinating mystery. He thumbed through his notes thoughtfully. *Yes, I really should write up an article on the global coincidences of drought.* Perhaps he could do that over the next few days—according to the sailor stories he had heard the previous night, the *Beagle*'s next stop didn't appear all that interesting. It was simply an unremarkable cluster of isolated volcanic islands in the middle of the Pacific Ocean. One wouldn't expect many spectacular findings at such a remote place.

A knock rapped on the wooden door of his cabin.

"Mr. Darwin, sir," the young voice of one of the young ensigns rasped unevenly. "The captain requests your presence on deck—our next port-of-call, the Galapagos Islands, has been sighted!"

Chapter 15

THE MYSTERY OF TIBET'S WEATHER SPIES

Many modern mystery books have elements of espionage in them. Yet few people—with the possible exception of a few conspiracy buffs who believe that the government is conducting secret weather-control experiments—would likely consider that a climate and weather book, even one about the world's greatest weather mysteries, could contain stories of amazing espionage. Nevertheless, at many times in the past, the clandestine acquisition of detailed knowledge of a region's weather was a matter of literal life or death. Surprisingly, real-life weather spies with the ingenuity of a James Bond were deployed to ascertain vital geographical and climatic information about one of the most inaccessible places in the world: the mountains and deserts of Central Asia.

Time: January 11, 1865
Location: Yarkund, near modern-day Yecheng, western Tibet in China, Asia
(This vignette is loosely based on two articles written to the Royal Geographical Society by Captain Thomas George Montgomerie in 1866 and 1868.)[1]

Sidi briskly rubbed his hands together in a somewhat futile attempt to warm them. Even under the assault of the bitterly cold wind whipping through the mountain pass, no complaint escaped the lips of this dusty-skinned man from India. Instead, he continued to walk up the dirt trail with an almost mechanical precision, clipping off strides of an astoundingly exact thirty-three inches, and periodically rolling his prayer beads in his numb fingers as if accounting for the pace.

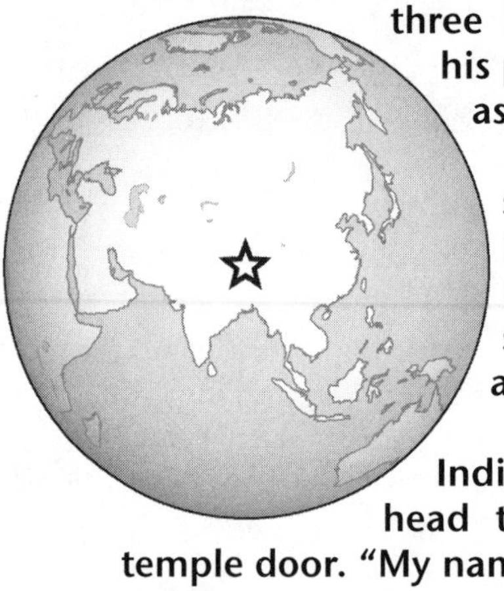

By nightfall, he reached a small alpine monastery nestled in the crags of the lofty mountains. Without hesitation, he strode up to the temple steps and knocked on the door.

"I am a weary traveler from India," he explained with lowered head to the monk who opened the temple door. "My name is Sidi. I would humbly ask for the hospitality of your temple on this cold night."

The monk inspected him carefully as the bitter north wind swirled through the partially open temple door. "You are welcome to rest here for the night," the holy man finally replied gravely. "Travelers may bed in the guest house." He gestured to a small stone hut across the courtyard.

Sidi nodded his thanks to the monk and walked to the stone shelter. He stuck his head inside, then glanced back toward the monastery to see if the priest was watching him—as, indeed, the man was. As Sidi knelt and entered the hut, he found it devoid of furni-

Mystery of Humanity's "Near Extinction." The author working on an Antarctic ice core drill. Photograph taken at Siple Station, Antarctica. *(Photograph courtesy of the author.)*

Mystery of the Saharan Hippos. Rock art image of a hippopotamus from the Wadi Mathendush, central Sahara Desert in Libya, Africa. *(Photograph courtesy of Dr. Andrea Zerboni of the University of Milan, Italy.)*

Mystery of the Vanishing Harappans. Drs. Pat Fall (Arizona State University) *(right)* and Caroline (Molly) Davies (University of Missouri at Kansas City) examine a dry lake sediment core from the Lisan Peninsula, Jordan. *(Photograph courtesy of Patricia Fall and Caroline Davies.)*

Mystery of Petra, the Rose City. The Treasury Building at Petra. *(Photograph courtesy of Dr. Tom Paradise of the University of Arkansas.)*

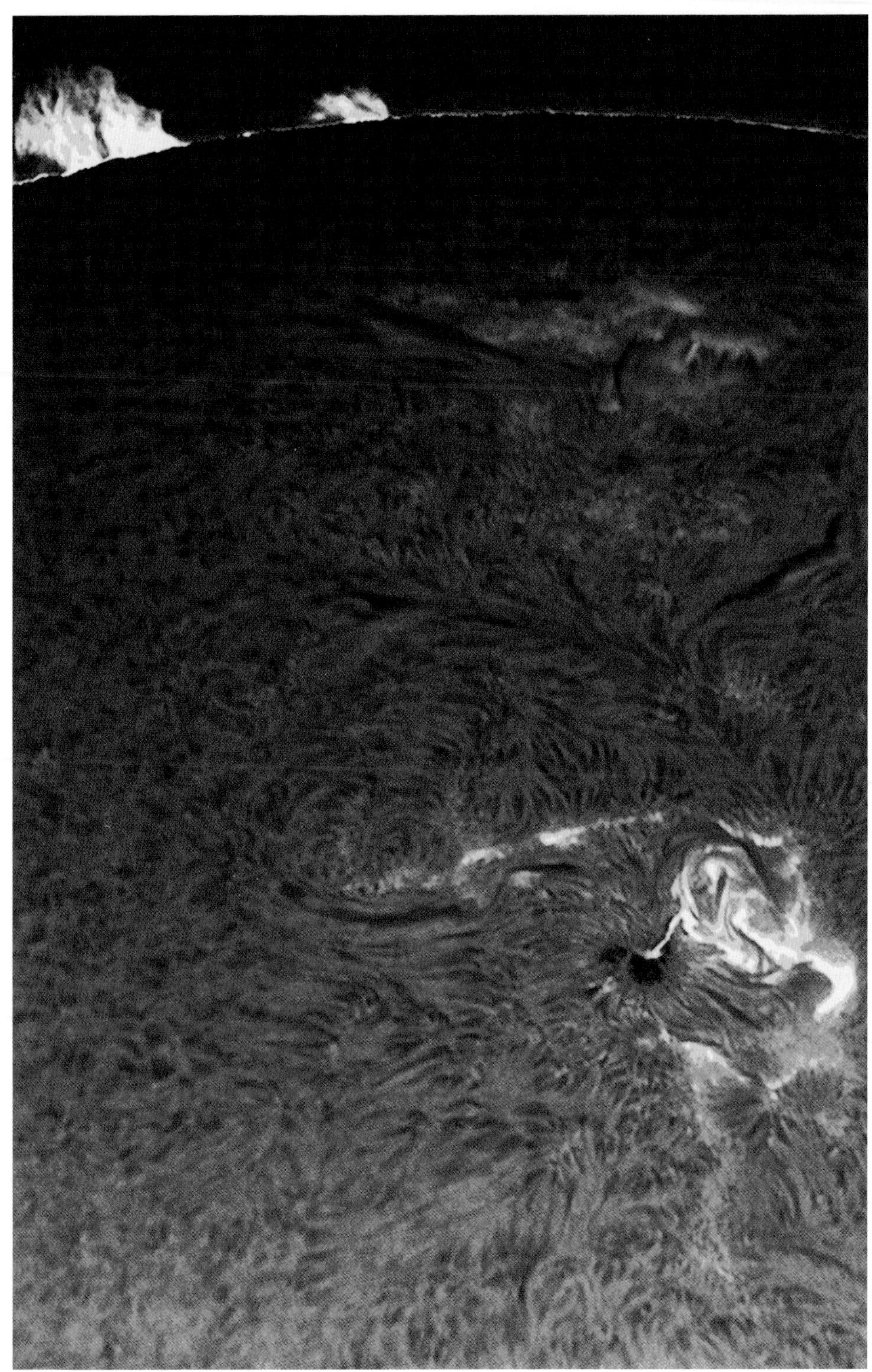

Mystery of the Little Ice Age's Lost Sunspots. Solar flare activity together with a sunspot *(lower right)*. *(Photograph courtesy of Greg Piepol, www.sungazer.net.)*

Mystery of the Mayan Megadrought. A Mayan temple in the Copan Ruins in Honduras. *(Photograph courtesy of Dr. Beth Larson-Keagy of Arizona State University.)*

Mystery of the Complacent Empire's Wet Weather. The summer palace, one of the residences of the Qing dynasty emperors, near Beijing, China. *(Photograph courtesy of Dr. Michael Kuby of Arizona State University.)*

Mystery of Tibet's Weather Spies. Geographer Mark Hildebrandt leading a trek through the still-mysterious Himalayas. *(Photograph courtesy of Dr. Mark Hildebrandt of the University of Illinois–Evanston.)*

Mystery of the Pacific Hot Tub. A sight that Charles Darwin likely experienced: a Galapagos lizard. *(Photograph courtesy of Dr. Michael Kuby of Arizona State University.)*

Mystery of the Devil in the Deep Blue Sea. One of the massive Greenland icebergs whose ancestors may have helped to create the Young Dryas period. *(Photograph courtesy of Dr. Merlin Lawson of the University of Nebraska–Lincoln.)*

Mystery of the Crashing Airplane and the A-Bomb. One of Fujita's air bombs: a microburst created by a desert thunderstorm. *(Photograph courtesy of Ron Holle.)*

Mystery of Climate, Cocaine, and the Storyteller. Drying coca leaves on a small Bolivian farm in the 1960s. *(Photograph courtesy of Dr. Ray Henkel, emeritus professor at Arizona State University.)*

ture except for a small pallet of straw at the far end that obviously was to be used as a bed. It would do, Sidi decided. He had slept in much worse. At the very least, it provided shelter against the bitter cold wind. However, before he could enjoy the relative warmth of the Tibetan guesthouse, he needed to conduct his mission.

Sidi peeked around the low-cut doorway of the hut. The Buddhist monk had shut the monastery door and returned to the warmth of the temple. Good, he could begin his real work.

Quietly, the Indian spy scurried outside of the guesthouse to the far side of the hut and carefully laid down his small traveling pack and his trekking pole. Glancing around to ensure his continued privacy, he grasped his walking stick and, with a quick twisting motion, unscrewed the metal-tipped bottom from the pole. Carefully, Sidi extracted a small, but high-quality, thermometer from the interior of the hollowed-out pole. Replacing the butt to the pole, he pressed the walking staff into the hard-packed ground and then hung the thermometer from it.

He next took his prayer wheel and carefully unscrewed the top portion containing the revolving cylinder from the shaft. Inside were secreted a number of tiny strips of parchment and a minute piece of lead. He meticulously examined the thermometer and, grasping the lead bit, wrote painstakingly the number of strides he had walked that day, 15,122 paces, then the temperature observation of 0°F, and finally the date of the observation. Darkness was falling quickly, Sidi realized as he glanced at the strikingly red mountaintops to the west. He hurriedly gathered a small pile of rocks, removed his pilgrim's beggar bowl from his pack, and positioned the bowl on the rocks.

From one of the lowland's greenish blue seashells (which, because of their rarity in this mountainous land, also served as a form of money), he painstakingly poured a tiny quantity of a thick, silvery liquid. The shell made an excellent container for the strange liquid

that the Sahib captain called "mercury." When the silvery substance had settled into the bottom of the bowl, Sidi used it to determine the precise geographic horizon—visually a very difficult operation in this mountainous terrain.

Finally, Sidi lifted his traveling pack, unfastened a tiny clasp, and opened the pack's false bottom. If discovered with this foreign—and therefore forbidden—instrument, Sidi thought soberly as he pulled the sextant from the case, he would undoubtedly be killed as a spy.

With the horizon set by the mercury in the begging bowl, he focused on the faint North Star through the deepening twilight. It was getting too dark to risk his lantern again—last night he had almost lost the light to an overly curious local official. He took the reading and then placed the sextant with the North Star reading still aligned carefully back into his pack. He wouldn't risk reading the device till the morrow's light.

All was done. Sidi reversed his operations—replacing the thermometer into the walking staff, returning the tiny paper strips and the lead writing tool back into the prayer wheel, and reclasping the tiny secret latch on his pack. The entire operation had taken only a couple of moments. His clandestine work for the British empire was, for the time being, completed, and he could sleep a few hours before returning to his important spy mission in the morning.

For over two and a half centuries, the remote mountains and deserts of Central Asia have been the playing grounds for an always clandestine, sometimes violent, and often deadly game of intrigue and political maneuvering between a collection of the world's great powers, which have included Russia, Great Britain, and the United States. Yet, on the whole, these covert activities in Central Asia during the last two centuries have been largely ignored by the rest of the world. The region's strategic importance was recognized briefly in 1901, when this strange superpower dance for control of Central Asia was given its most famous name—*the Great Game*—by the renowned Rudyard Kipling.[2] Currently, as the secrets of many of these countries are becoming public for the first time, we are now beginning to realize that the many stories and adventures of the Great Game comprise some of the greatest spy and espionage exploits ever recorded—and, interestingly, weather has played an integral role in many of those adventures. Indeed, the entire political face of the world today might have been very different were it not for the Great Game's weather.

It all began in the early nineteenth century when concern began to grow in Great Britain and British-controlled India over the expansionist plans of the Russian tsars and of Napoleon Bonaparte. Although the British navy protected the ocean access to the riches of India, the de facto rulers of India, the directors of the East India Company, feared a disturbing potential existed for invasion from the north. Yet little was known of the mountainous interior regions of Asia. Strict Muslim control and an inbred distrust and fear of foreigners meant that Europeans were as likely to be beheaded as welcomed if discovered in these lands.

Yet, despite the grave risks associated with travel in these areas, the climate and the terrain of Asia's remote interior had to be ascertained for the future military needs of the great powers. And that need created the opportunity for some of the world's best spies to develop their craft. Ambitious military officers among the Russians and from the British undertook secret missions into the Himalayas and Afghanistan. They covertly recorded information on the geographic nature of the passes, the characteristics and society of the region's peoples—and the weather.

One of the first participants in the Great Game was a young British military officer named Henry Pottinger. In 1810, his superiors gave the twenty-year-old lieutenant of the Fifth Bombay Native Infantry the extremely dangerous mission of conducting a comprehensive reconnaissance through the unknown regions northwest of India. The young subaltern wisely decided to disguise himself as an Asiatic Muslim and adopted the role of a horse trader. With almost no knowledge of what he was about to experience, and attempting a route from northwest India to Persia that no other European would travel for another century, Pottinger set out from the village of Nushki in March of 1810. His journey to Persia (present-day Iran) took three months and covered hundreds of miles across two scorching deserts.

His first unusual geographic encounter was with the huge sand dunes of the expansive Helmund Desert. "Most of these," he recounted, "rise perpendicularly on the opposite side to that from which the prevailing wind blows . . . and might readily be fancied, at a distance, to resemble a new brick wall."[3] Yet surmounting these twenty-foot dunes was "trifling compared with the distress suffered, not only by myself and people, but even the camels, from the floating particles of sand."[4] The suspended layer of abrasive red dust, Pottinger recorded, often got into his eyes, nose, and mouth, creating extreme discomfort and adding to his growing thirst in the intense heat of the desert.

Although a major desert storm, or *haboob*, brought Pottinger and his local guides short-lived relief from the heat, it was quickly replaced by a desert downpour of rain. "The rain fell in the largest drops I ever remember to have seen, and the air was so completely darkened, that I was absolutely unable to discern anything at the distance of even five yards."[5] The young soldier's guide told Pottinger that during the summer (the Briton was traveling in April) the winds could become deadly. "The winds in this desert are often so scorching and destructive, as to kill any thing, either animal or vegetable, that may be exposed to them."[6] The effects of these winds as described by Potttinger's guide were horrific: "The muscles of the unhappy sufferer become rigid and contracted; the skin shrivels, and agonizing sensation, as if the flesh was on fire . . . and in the last stage, it cracks into deep gashes, producing hemorrhage, that quickly ends this misery."[7] Fortunately, Pottinger managed to avoid such killer winds during his travels.

To gain entry to the remote regions of Persia, the young British spy proved inventive, employing his horse trader disguise. For example, following one meal at a small desert village, the village head requested that the disguised soldier give "a Fatifu or prayer of thankgiving."[8] Pottinger recounted, "This was as unexpected as it was unwelcome, and I was greatly perplexed for an instant."[9] Yet he rose to the task. "I assumed a very grave air, stroked down my beard with all imaginable significance, and muttered a few sentences, managing to articulate rather distinctly the words Ullah [Allah], Rasool [the Prophet], and Shookr [Thanks]."[10] Apparently the lieutenant's deception was successful, for the locals were satisfied and allowed him to continue on his way.

Unfortunately, the amenities of local villages were not always present during his trek. Once, as he crossed a large barren expanse of desert in southern Persia (present-day Iran), Pottinger recounted that the heat was "greater and more oppressive than I had hitherto experienced since leaving India."[11] He noted that a person may endure such misfortunes as "the pressure of fatigue or hunger, heat or cold and even a total deprivation of natural rest for a considerable length of time."[12] "But," young Pottinger continued, "to be scorched under a burning sun, to feel your throat so parched and dry that you respire with difficulty, to dread moving your tongue in your mouth from the apprehensions of suffocation it causes, and not have the means of allaying those dreadful sensations, are in my ideas the extreme pitch of a traveller's calamities."[13]

Eventually and amazingly, Pottinger reached the shah's palace in Persia and completed his mission. The trip had proven his worth to king and country; Pottinger eventually rose through the ranks of British bureaucracy to become the first governor of Hong Kong. Although Pottinger was successful in his travels through the inhospitable regions of Central Asia, others were not quite as lucky.

When the British had begun their secret reconnaissance of the unknown area of Central Asia, the Russians had not remained idle. In 1839, they assembled a huge five-thousand-man army, with a supply train of ten thousand camels carrying food, guns, and ammunition, and set out from Orenburg in southern Russia toward the strategically important city of Khiva (now in the present-day country of Uzbekistan). Initially the army had been told that they were to be "an escort" for a "science mission to the Aral Sea."[14] Later, it was revealed that they were to be an occupying force for the expansionist Russian empire's movement southward.

At first, the army met no resistance. The expedition had been planned to begin in the early winter to avoid the terrible heat that had been so meticulously detailed by early explorers such as Pottinger. The army's leader, General Perovsky, had hoped to arrive at Khiva before the worst of the Asian winter set in. Nevertheless, the army hit increasingly bitter winter conditions. The official expedition report noted that at night in their felt tents, the Russians were forced to cover their entire bodies with sheepskin coats to protect their extremities, particularly their noses, from frostbite.

Although the cold might have been expected, the snow was not.

As the expedition continued southward through December, more and more snow fell. Soon, even the local guides began to remark that such snow was unprecedented. The tracks for the army began to quickly fade as the snow continued to fall. The report noted that "beyond a distance of 20 yards, no object was to be seen through the clouds of snow which were whirled about in every direction. The fury of the storm was so great that it was impossible to draw breath when facing the wind, and the intense cold penetrated to the bones."[15]

Camels began to die at an alarmingly rate. "Once a camel fell," the official report stated, "it rarely rose again."[16] By early January, almost half of the supply train animals had died and the surviving pack camels, crazed by hunger, began to turn their heads and gnaw through their cargo of wooden ration crates. That forced the army to have to unload the thousands of ration boxes off the camels every night and then agonizingly reload them every morning, in addition to the normal camp-making procedures. The official report noted that only by eight or nine in the evening could the soldiers obtain a little rest.[17] But then "the soldiers of the detachment were generally astir at about 2 A.M., when preparations were commenced by the cook and servants for boiling tea and buck-wheat porridge. On some days however, notwithstanding the severity of the frost, the detachment had to commence its march after a repast of frozen biscuits in consequence of there being no fuel to light fires."[18]

The snow continued to fall. The drifts were now so deep that the men with snow up to their waists were required to forge a path for the camels and artillery. Morale plummeted and reports of illness increased. By January, two hundred men had died of sickness and four hundred more suffered from debilitating illnesses. Yet, surprisingly, the official report specifically noted, "Not a single man had died from cold, although there were numerous cases of frost bites."[19] Camels died at the rate of a hundred per day. By this time they had traversed less than half the distance to Khiva. And scouts were reporting that the worst still lay ahead.

After discussion with his column commanders, General Perovsky decided to abandon the invasion and called for retreat. Unfortunately, the retreat was to be as hazardous as the initial advance. In addition to the snow, the freezing winds, and the sicknesses, the trail of dead camels left by the army had drawn packs of wolves that began to attack the retreating columns of men and animals. The official report stated, "the route pursued by the columns in front could be ascertained by the pillars of snow erected at some distance from each other by the Ural Cossacks, by the snow heaps that marked the night camps, and by the camels, living and dead, some frozen and partly devoured by wild beasts, that lay along the line of march."[20] The continual snow, coupled with the poor diet of the army, led to snow blindness.

When what remained of the army finally staggered back to Orenburg in May of 1840, the final numbers starkly demonstrated the magnitude of the disaster. Of the original five-thousand-man army, over a thousand had perished in the bleak interior deserts. Only fifteen hundred camels of the original ten-thousand-supply caravan survived to return to Orenburg. The southeast expansion of the Russians was halted—at least for a time.

Eventually the Russians overcame their devastating failure and in the 1860s commenced a measured conquest of the Muslim khanates of Khiva, Bokhara, and Khokand (which today form the countries of Uzbekistan, Turkmenistan, and Tajikistan). The British in India grew increasingly alarmed at the Russian encroachment southward. Could the Russians be considering a move through Tibet into India? And, if so, exactly what route would their armies travel? These critical questions could not be immediately answered. At that time, very little was known of the eastern Himalayas. Were there traversable passes? What were the distances involved? Would weather conditions be tolerable?

Obviously, more detailed information on this huge mysterious area was needed. Yet, as Pottinger and other less fortunate spies had learned, foreigners were seldom welcomed and were often subjected to grave danger in these remote regions. A British officer, Captain Thomas George Montgomerie, came upon a brilliant solution. He wrote, "When I was in Ladakh I noticed the natives of India passed freely backwards and forwards between Ladakh and Yarkund in Chinese Turkestan, and it consequently occurred to me that it might be possible to make the exploration by that means. If a sharp enough man could be found, he would have no difficulty in carrying a few instruments amongst his merchandise, and with their aid good service might be rendered to geography."[21]

So Montgomerie was given permission to create an ingenious spy ring of native explorers. All were trained in clandestine surveying techniques and selected for their intelligence and adaptability. These spies, whose real names were not revealed until they died, were known as the *Pundits*, from a Sanskrit term referring to a learned man. Before beginning their missions, the Pundits were put through what today might be termed a spy school. There they extensively practiced such exotic techniques as trekking with an exact stride (in one case, recorded as precisely thirty-three inches per pace) and how to keep daily tally of their steps by using Tibetan prayer beads.

They also were given spy equipment that the fictional "Q" of the James Bond stories would have been proud of. Their Buddhist prayer wheels were constructed with secret compartments in which tiny scrolls of paper could be concealed for marking routes or noting conditions along a given mountain trail. A sextant was secreted in the false bottom of their traveling chests. Mercury was the substance needed for setting an artificial horizon when taking sextant readings. It could be hidden in one of the sealed multicolored seashells that were used as money in some of the more remote parts of Asia, and then poured into a pilgrim's bowl when needed for readings.

A thermometer was hidden in a hollowed-out walking stave. This device provided a dual function. One purpose of the thermometers was, of course, to determine the weather conditions along the trek. Second, and perhaps more important, the thermometer was instrumental in ascertaining altitude, very critical information for military analysts. But how can one measure altitude with a thermometer?

Because of the normal decrease of barometric pressure with height, the boiling point of water decreases 1.8°F for each thousand-foot increase of altitude. Consequently, the Pundits used their thermometers as *hypsometers*, devices that measure the water's boiling point to determine the actual elevation. Indeed, the official inventory of spy equipment given to each Pundit included "1 Copper jug and oil lamp for boiling the thermometers."[22] But was the method accurate?

One of the first measurements made by the Pundits was at the high-elevation village of Leh at 11,278 feet above sea level. The Pundit found the "temperature of the air 42°[F] at noon, and water boiled at 181°3 of Fahrenheit."[23] In another instance using their thermometers, the Pundits calculated an altitude of 11,700 feet above sea level at the Tibetan capital of Lhasa. The current accepted elevation for the capital is given as 12,000 feet above sea level. Remarkably, in evaluating the Pundit's estimate of the Tibetan city's elevation, the spymaster Montgomerie himself judged that it was probably "only four or five hundred feet higher."[24]

The weather of the unexplored region was not neglected by the Pundit spies. As I have fictionalized at the beginning of this mystery, the spymaster Montgomerie wrote in an official report that the Pundits found severe weather in Yarkund during the winter of 1864–65: "The thermometer early in January having fallen nearly to zero [Fahrenheit], or 32° below the freezing-point. At times the weather was cloudy, and from the 19th to the 26th January snow fell; but, judging from the general regularity of the observations of the whole must have been very clear."[25] The diligence of the Pundit in recording the weather conditions at Lhasa and other locations was remarkable. The Pundit recorded an amazing 259 separate hourly observations from February 9 to March 9, 1866.

The success of the Pundit spies is undisputed. Strangely, however, one of the most bizarre aspects of this whole affair was how the integrity of the operation was almost completely compromised by its spymaster, Captain Montgomerie. In 1866, immediately after one of the first successful missions, Montgomerie wrote a comprehensive report on the undertaking (including aspects of the Pundits' disguises and discussion on some of their secret equipment) for the respected *Journal of the Royal Geographical Society*. In this article, he even hinted as to the real purpose of the Pundits' travels: "The progress of Russia in the Ileh valley seems to be correctly noted; but whether he [the Pundit] is right that the Russians have a fort near Lake Lop . . . is very doubtful."[26]

After that first security breach, Montgomerie submitted a second report in 1868 to the journal, in which he gave complete details on the Pundits' illicit spy equipment on a mission to Lhasa. He wrote, "Reading the sextant at night without exciting remark was by no means easy. At first a common bull's-eye lantern answered capitally, but it was seen and admired by . . . curious officials . . . and the Pundit . . . was forced to part with it in order to avoid suspicion. [Consequently,] the Pundit was at some of the smaller places obliged to take his night observation and then put his instrument carefully by, and not read it till the next morning."[27]

Discussion of the meteorology of the region also appeared in the second report: "During the whole time the Pundit was in the Lhasa territory, from September to the end of June, it never rained, and snow only fell once whilst he was on the march, and twice whilst in Lhasa. The snow-fall at Shigátze was said to be never more than 12 inches; but the cold in the open air must have been intense as the water of running streams freezes if the current is not very strong. A good deal of rain falls during July and August about Shigátze, and there is said to be a little lightning and thunder, but the Pundit does not recollect seeing the one or hearing the other whilst he was in the Lhasa territory."[28]

This massive breach of security (and obvious endangerment of the Pundits who were still operating in Tibet) is hard to understand. Captain Montgomerie did make a small disclaimer that "the two Pundits, being still employed on explorations, their names are, for obvious reasons, omitted."[29] In Montgomerie's defense, it should be noted that the *Journal of the Royal Geographical Society* was not for public sale (as it was only distributed to members of the society). However, the society was international—and its membership included Russians. It is likely that the tsarist agents in Central Asia found Montgomerie's detailed reports to be invaluable sources of information.

Is the Great Game still played today?

Although occasionally the world players have changed, such as with the collapse of the Soviet Union, the remote territories of Central Asia remain critical—if not always public—regions of world concern. And heroic participants still exist in today's version of the Great Game.

I have had the great fortune to work with one of the best. The late Melvin G. Marcus was a premier mountaineer and meteorologist of the likes of Lieutenant Pottinger or the Pundit spies of Tibet. Indeed, Marcus, a virtual "mountain of a man" and respected professor of geography, is credited with making a number of first ascents—the first to reach the summit of a given mountain—and was friends with such great alpinists as Heinrich Harrer. Harrer was a teacher of the Dalai Lama and the author of the fascinating book *Seven Years in Tibet*. Marcus's hands-on field measurements in such remote and diverse locations as the bottom of the Grand Canyon, atop glaciers in Alaska, or even astride a donkey in the Himalayas, have proven useful to academic institutions—and undoubtedly other less public organizations—around the world.[30] And his students, whom Marcus trained throughout his long career, continue to carry out on-site geographical fieldwork.

The late alpine geographer Mel Marcus setting up a remote weather station in the Himalayas. Photograph courtesy of Dr. Anthony Brazel of Arizona State University.

We also now have highly sophisticated equipment in fill in the gaps left by the field geographers who journey to those remote lands of the Himalayas in pursuit of on-site knowledge. Satellite observing platforms, for example, now allow analysts an astonishingly detailed view of the inaccessible lands of Central Asia. So we continue to eke out valuable nuggets of knowledge about that last great hidden region of the world as we seek to discover its multitude of secrets.

Time: Spring 2013
Location: Central Intelligence Agency Headquarters, Langley, Virginia, North America

"Bring up the current infrared satellite image of that storm over Tibet," the intelligence chief ordered.

The plasma screen flickered to reveal a massive white splotch of cloudiness centered over the Himalayas.

"Can we have an overlay of the upper air wind patterns put over that?" the chief then requested.

With a few keystrokes, the analyst deftly updated the image with a series of flowing lines indicating the high-altitude wind patterns for the area.

"Given the speed and direction of those winds, that blasted storm should move that way," the chief muttered, gesturing to the upper-right corner of the screen.

The analyst nodded and replaced the image with the results of a computer-generated weather forecast model. As the chief had suggested, the computer model indicated a future northeastward movement to the present storm.

"Looks like it will clear the area in thirty-six hours, sir."

"Good, then we can advise the president that his initiative to open relations between the Dalai Lama with the Chinese leadership can continue with a face-to-face meeting in Lhasa. The storm will pass by the time of the meeting's start. For that meeting, at least, weather won't be an issue—although, I imagine, pretty much everything else will be . . ."

Chapter 16

THE MYSTERY OF THE GREAT AMERICAN DUST BOWL

As spies traveled through the inaccessible Himalayas during the last two centuries, more settled parts of the world were carefully recording the weather. We now progress into more recent periods of history, armed with more knowledge about the weather and climate. For example, our understanding of the Dust Bowl days of the 1930s is based on a huge variety of information ranging from such poignant folk songs as "So Long, It's Been Good to Know You" by Woody Guthrie, to the stark black-and-white photographs of wind-hardened, gaunt men and women, to even the detailed weather logs meticulously handprinted by weather bureau employees. Yet even with all of that information, the underlying causes of that Great Plains drought remain a mystery even today.

Throughout this mystery, we'll discover that some of the critical "green" issues of today's climate debate were being discussed (and actually implemented by Congress with intriguing results) over a hundred and fifty years ago. And we'll learn that the sun and moon, along with people's misapplied environmental actions, may have been the prime causes of the great Dust Bowl droughts of the 1930s.

Time: April 14, 1935
Location: A small community Methodist church in the western panhandle town of Guymon, Oklahoma, North America
(This vignette is loosely based on a famous sentence from an article written on April 15, 1935, by Associated Press reporter Robert Geiger.)

The reporter surreptitiously inspected the small church's congregation with an adept eye. It was as if the ever-present dust of this now godforsaken land was literally embedding itself into the clothes and even the skins of these poor people. Everything—windows, pews, and the parishioners themselves—was tinged a dingy brown hue. Acrid dust hung in the air, visible as dancing specks in the subdued weak sunlight filtering through the ubiquitous russet haze.

Numerous dry hacking coughs broke the contemplative silence of the minister's prayer.

The reporter pulled out his small notebook and jotted down a few lines. "Consumption and pneumonia, no doubt aided by the voluminous quantities of dust in the air, are reaping a grim harvest of these struggling farmers and their starving families."

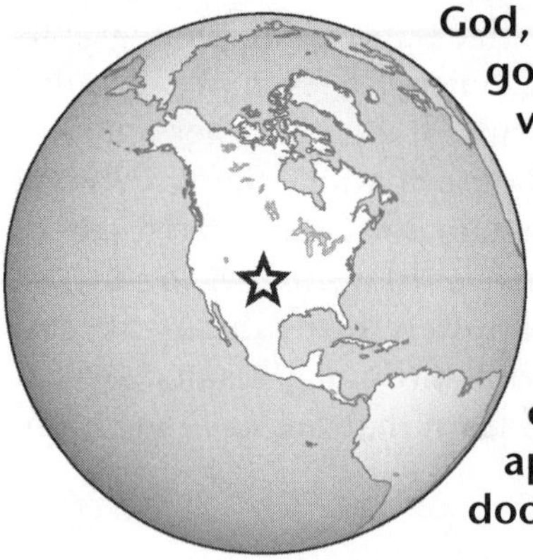

"Please, oh compassionate and caring God, deliver thy precious rain to these good people . . . ," the minister fervently preached through the subdued cacophony of hacking coughs.

With a startling bang, the doors of the church slammed open as if jerked by a dust-covered demon. A skeletal-appearing man appeared in the doorway. He was perhaps only in his fifties but his leathery, wind-worn dusty features suggested a much older age.

"Sorry to stop your fine speechifying, Pastor," the grizzled man called from the opening. Behind the weather-beaten man, the reporter noted an ominous "emptiness" to the afternoon air. No sunlight filtered through the doorway as might be expected in the open plains, only an opaque darkness surrounded the man. He called out, "There's one devil of a dust storm abearin' right down on us! You'd all had a better take cover!"

The congregation quickly responded by shuffling toward the back of the church—no panic, the reporter noted, it was as if this were simply one more obstacle that some malevolent god or devil was employing to test these besieged people. The pastor was telling people to head into the interior coatroom at the back of the church. A swelling moan of wind combined with the thick, deepening darkness made it appear as though the church were caught in the gravelly maul of the Devil himself.

The dust blizzard continued to batter the church. A small child—perhaps only five or six years old—looked up into his mother's eyes. "Will it ever end, Mommy?"

The mother pulled the child close to her and hugged him tightly. The reporter was close enough that he could make out the raspy words she spoke to soothe the youngster.

"If it rains, little one," she cooed softly, "if it rains."

The reporter licked the end of his pencil, already tasting as much dust as lead on its tip.

"Three little words achingly familiar on the Western farmer's tongue, rule life in this dust bowl of the continent . . ." He paused in his writing, looking over the huddled drought-blasted mass of humanity in the cloakroom as if searching for just the right phrase to capture this dreary brown landscape.

"If it rains."

The infamous "Black Blizzard" dust storm of Sunday, April 14, 1935, was a potent symbol of how bad a drought can be. Although the day had dawned clear and beautiful across the high plains stretching from Texas to Kansas, toiling farmers saw an ominous black cloud appear on the northwestern horizon. Daylight was abruptly transformed into utter darkness as gale-force winds ripped millions of tons of black topsoil from the ground and lifted it thousands of feet into the air—then hurled the fertile soil thousands of miles away from those despairing farmers. Indeed, the massive Black Blizzard of April 14, 1935, actually blew fine reddish dust from the Oklahoma prairie all the way to Washington, DC—tangible evidence to the country's politicians of the horrible plight of the Great Plains' farmers half a continent away.

The American Dust Bowl of the 1930s was the result of an interesting juxtaposition of both climatic and social factors. One of the amplifying aspects of the 1930s drought was the traditional farming practices by the people of the area. Those cultivation patterns and methods of farming in that region could be traced back to the initial settlement of the Great Plains in the 1800s. At that time, little was known of the Great Plains' climate. Although several expeditions had explored the region in the early 1800s, they had not been evaluating the region for its agricultural potential. Furthermore, their findings were sometimes buried in government reports that were not readily available to the general public.

Misleading information, however, was plentiful. Land speculators for the region, hoping to promote settlement, put forth glowing but often wildly inaccurate accounts of the Great Plains' agricultural potential. In addition to this false information, most settlers had little money or other assets for converting the land to agriculture, having spent much of their pocketbooks to get to the region. Moreover, their farming experiences were completely based on conditions in the more humid eastern United States. Consequently, the crops that they selected to grow and the cultivation practices that they used were often not suitable for the Great Plains. Not surprisingly, after many failed attempts to grow eastern crops, some of these people—and even some government and military officials—labeled the Great Plains the Great American Desert.[1]

In the 1970s, Dr. Merlin Lawson—my mentor at the University of Nebraska and a noted historical climatologist—broached an interesting idea about the weather of those Oregon Trail days to the talented Arizona dendrochronologist Charles Stockton. The two began to critically examine the historical accuracy of that "desert" label for the American Midwest. They determined through an exhaustive search and analysis of tree ring records, military post weather records, and even Oregon Trail diaries and letters, that the Great Plains was *not* abnormally dry in that initial settlement period of the first half of the nineteenth century. Indeed, their research suggests that the early 1800s may have been actually wetter than the present. Yet compared to the East from where the settlers originated, it was much drier.

As many people do today, those early settlers of the Great Plains, rather than questioning their own conceptions of the region's environment as well as their farming practices, believed that the climate of the region was changing. Throughout the early 1800s, a very spirited debate (both scientific and political) had begun on whether the

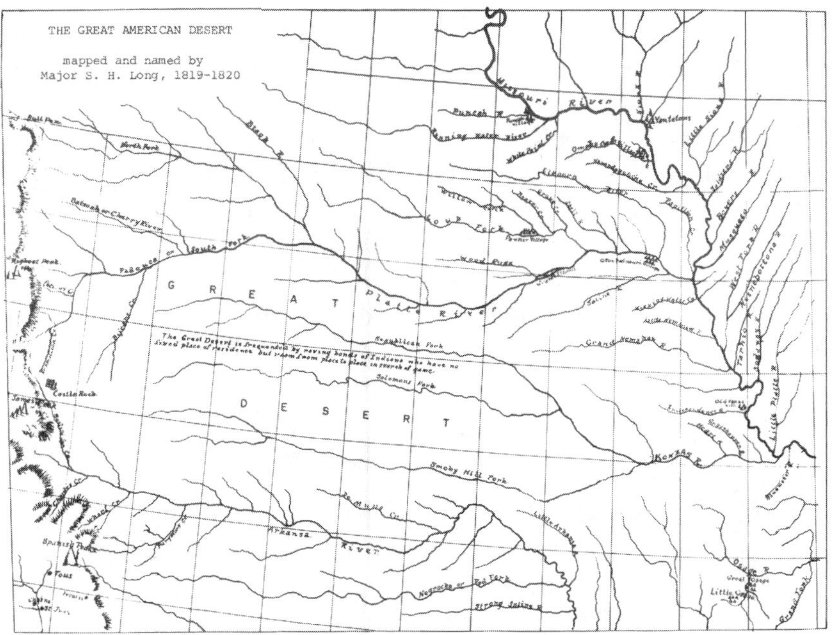

One of the first maps dating to the 1820s that identified the American Great Plains as a "Great Desert." Map courtesy of Dr. Merlin Lawson of the University of Nebraska.

world's (or least the country's) climate was changing, and (relevant to today's world) whether those changes were the direct result of human activities.

For example, in the early 1800s Dr. Noah Webster (of dictionary fame) made a "most extensive review" of the climate change question for both North America and Europe.[2] From that review, he arrived at the conclusion that "the hypothesis of a moderation of climate appears to be unsupported" for either continent.[3] Yet, the respected magazine *Scientific American* published a letter in 1869 stating, "The scouts, guides and hunters all agree in stating that on the Plains, as far back as their experience goes, little or no rains have fallen during the summer. . . . Has the iron of the rails or the upturned ground the credit of the change?"[4]

Unsurprisingly, it was this latter view that won the attention of the legislators (who were, of course, responding to their increasingly disgruntled voters). So, these politicians (and some scientists) reasoned that, if humans were causing a lack of rainfall in the region by building railroads and plowing up the soil, then people could also fix the problem by changing climate. For instance, one Kansas scientist in 1872 wrote that "if every farmer settling upon the high prairies would each year put out only a small piece of thickly planted forest . . . we might hope that . . . the climate of the whole country would be materially modified."[5] The fundamental reasoning was that, by planting trees, the region's humidity would increase and rains would become more prevalent.

So, to encourage tree planting in the region (and thereby, hopefully change climate), Congress passed the Timber Culture Act in 1873, mandating that a settler could homestead an additional 160 acres of land under the stipulation that he allocate 40 acres of that land entirely to tree cultivation. Over the next couple of decades in the 1800s, some 10 million acres of the Great Plains were allocated under this act. But did it work?

Unfortunately, as Nebraska geographer Barron McIntosh discovered in his research a century later, it did not—for two reasons, one social and one climatic.[6] First, the social problem: not all people acted in the "green" spirit of the Timber Culture Act or in the best interests of society in general. The act generated a massive amount of speculative land fraud because of an exploited loophole. To claim land under the act, one didn't actually have to live on the land. For example, an inspector for the act wrote in 1885 that a "vicious system of fraudulent entries has been successfully practiced by and in the interests of cattlemen and stock corporations. If the law had been enacted solely for their benefit it could scarcely have been more successful. . . . The method is simple, effective and infamous. A 'cattle king' employs a number of men . . . and each . . . is expected and required to make a timber culture entry."[7]

Second, there was a climatic problem. To grow trees, one needs a consistent and rather abundant amount of moisture. Other than along rivers (and even some of those tend to be periodically dry in the Great Plains), there simply wasn't (and still isn't) enough water consistently available (without deep-well irrigation) to grow large enough groves of trees to have any impact on the region's climate. It was truly a Catch-22 situation for the Timber Culture Act: in order to create more rain, one had to plant trees . . . but for the trees to grow, the area needed more rain.

In the face of both of those problems, social and climatic, the Timber Culture Act was repealed in 1891.

Oddly enough, however, the people who had noted that the weather by the end of the nineteenth century seemed different from that at the start of the century were actually right—and, as they had suggested, perhaps the settlers themselves were part of the cause. As we now know, the middle of the nineteenth century marked the end of the Little Ice Age—and the Little Ice Age's demise has been attributed both to a return to a normal sunspot cycle from the Maunder Minimum, as I discussed earlier, and to the atmospheric global warming associated with the Industrial Revolution.

And it is that juxtaposition of both solar activity and the actions of people that might have been the major contributing factors to the creation of the Dust Bowl. While the farming practices of the Great Plains settlers—and the increasing need to put more land under cultivation to pay for the equipment and seed—undoubtedly played a role in the severity of the Dust Bowl, it is possible that other factors contributed as well.

With the end of the Maunder Minimum and return of a strong, well-defined sunspot cycle by the mid-1800s, many scientists attempted to link sunspot activity to rainfall. Their physical argument was that more sunspots related to higher amounts of solar radiation, and this energy increase would intensify precipitation processes across the globe. By the 1870s (the same decade as the Timber Culture Act), some twenty or more scientific papers had been published relating rainfall in different parts of the world (India, Great Britain, and the United States, for example) to the periodic sunspot cycle. By the 1920s, with longer and more detailed rainfall records, the rainfall/sunspot theory had become a bit more complicated. H. H. Clayton, an American climatologist, suggested that some parts of the world might become wetter during the peak of a sunspot cycle but other parts might become drier, because the changes in solar radiation would cause pressure patterns, and their associated weather systems, to shift.[8]

Research into sunspot/drought relationships suffered a decline during World War II. But following that war, some scientists—including my wise mentor Merlin Lawson—noted a seeming periodicity for droughts. Researchers had recorded that a strong Great Plains drought had just occurred in the 1950s. They matched that drought to the Dust Bowl droughts in the 1930s and saw a difference of roughly twenty-two years—the same length as that of the Hale double-sunspot cycle. When the Great Plains experienced droughts again in the early 1980s (and again during the first years of the twenty-first century), it appeared to some people that the similarity in timing of droughts and sunspot activity was beyond coincidence.

The trouble with immediately connecting sunspot activity and drought occurrence through these three events is the very distinct possibility of statistical chance. Even though three droughts matching a set of sunspot cycles might appear beyond accidental, the random occurrence of three things in a sequence isn't unprecedented. Statisticians speak of *n*-size, the number of events that make up the sample. The larger the *n*-size is, the greater the probability that the linkage—if it is there—isn't simply random statistical chance.

But how do we increase the *n*-size? Charles Stockton and David Meko of the University of Arizona's Laboratory for Tree-Ring Research collected and collated tree ring samples across the western two-thirds of the United States. They used that multitude of tree ring samples to re-create the rainfall record of the western United States back to 1600 CE—and the *n*-size jumped from a relatively short century of rainfall records to a four-hundred-year database of reconstructed rainfall.[9]

Enter J. Murray Mitchell, a skilled climatologist from the

National Oceanic and Atmospheric Administration. Mitchell attempted to prove that periodic twenty-two-year cycles were indeed embedded in the four-hundred-year reconstructed rainfall record of the Great Plains and that those cycles corresponded to the sunspot cycle. So, working with the two Arizona tree ring researchers, Stockton and Meko, J. Murray Mitchell employed a powerful statistical tool called *spectral analysis*.[10]

Originally called *power spectral analysis*, spectral analysis was first developed by electrical engineers to identify repetitive frequencies in power and sound generation. Climatologists have applied the most basic function of spectral analysis to identify cycles or periodicities in studying various proxy records of climate, such as lake sediments, ice cores, or tree rings. Spectral analysis requires a long-time sequence of data—for example, the reconstructed temperature record over the past few hundred years from tree rings—and identifies any underlying cycles or periodicities in that sequence of years.

What Mitchell, Stockton, and Meko disentangled from their mass of data through spectral analysis was exactly what Merlin Lawson had anticipated way back at the beginning: a prominent repetitive cycle in rainfall every twenty-two years. But this wasn't enough for the researchers; they fine-tuned their analyses to concentrate on that twenty-two-year cycle and discovered that the similarity in timing between the tree ring–created rainfall record and the sunspot cycle was not the product of simple random fluctuations in the statistical analysis. Although those results didn't mandate any cause and effect (statistics can't do that), the periodic similarity in cycles of rainfall and solar activity was significant enough to warrant more study.

The researcher who followed up on their work was a professor from the New York State University at Stonybrook named Robert Currie.[11] Currie devoted most of his research to delving into climate records in search of solar and lunar cycles. I mentioned earlier that the sun demonstrates an eleven-year cycle in sunspot activity with a Hale double-sunspot cycle involving magnetic polarity reversal lasting twenty-two years. The moon also demonstrates strong repetitive cycles. One of the strongest of those is the 18.6-year *lunar declination cycle*. The moon's orbit swings north and south of the terrestrial equator so that it progressively moves more poleward (both north and south of the equator) and then back to oscillating closer to the equator over a regular repeatable period of 18.6 years. Given the moon's proximity, such variations cause long-term tidal changes on Earth. Currie believed that such tidal changes created variations in regional climates around the world.

Not being a climate reconstructionist like Stockton or Meko, who used climatic substitute data such as tree ring records, Currie instead unearthed massive quantities of historic actual weather records from around the world. He published over fifty professional research papers examining these weather records for solar and lunar cycles. Some might say he even went a bit overboard. For example, in one paper he identified an 18.6-year lunar cycle existing in 1,015 out 1,219 individual weather records.[12]

But he ran into a spot of trouble. Over the long course of his data gathering, he determined that the connection between rain frequency and lunar tides wasn't consistent in intensity, geographic pattern, or even in the specific phase of the lunar cycle.[13] For example, rainfall records from the Mid-Atlantic and New England states showed that maximum lunar tides produced dry periods in the nineteenth century in a repetitive pattern—but that pattern broke down by the late 1800s. Inexplicitly, the pattern then reestablished itself by 1917, again producing periodic dry periods with lunar maximal tides.

That presented a big problem for scientists: Were the lunar cycles or the solar cycles the primary causes of the periodic droughts in the American Midwest?

J. Murray Mitchell was one of the first climatologists to recognize that those Midwestern droughts might be influenced by both solar and lunar cycles. When he carefully studied the time period from 1600 to 1962 as a whole, Mitchell discovered that there was no indication of a lunar signal in the rainfall records obtained from the tree rings. But, if that time period was broken into two halves, then the lunar cycle appeared as strong as the solar cycle.

The reason for that seeming contradiction is that the lunar declination cycle abruptly shifted 180 degrees in phase around the year 1780. Consequently, if one was examining the entire time period, the lunar signal from the early record almost completely canceled out the signal from the late record. The original statistical work of Mitchell, Stockton, and Meko, inspired by Lawson's idea of drought periodicity, simply wasn't precise enough to extract that phase shift change from the data.[14]

Subsequent work in the late 1990s by Ed Cook, working with those diligent Arizona researchers Meko and Stockton, confirmed Mitchell's findings; there was a strong statistical relationship between a roughly twenty-year American Midwest drought cycle and both the Hale double-sunspot cycle of twenty-two years and the 18.6-year cycle of lunar tides.[15] This relationship appeared particularly strong after the year 1800.

So, if substantial research said drought is linked to solar and lunar cycles, why aren't we using this relationship to forecast future droughts?

Scientists like cause and effect. Rather than just say that variations in one aspect of nature match variations in another aspect, scientists in general like to explain through accepted physical processes why the association exists. With the solar-lunar cycles and drought, we simply can't say if the drought is the result of the combined solar-lunar cycles, or something else.

For example, researchers have recently identified an ocean circulation in the North Pacific Ocean that appears to operate on cycles of forty to sixty years. That circulation is called the Pacific Decadal Oscillation (PDO).[16] PDO variations are manifested as changes in surface temperature and pressure over the North Pacific. During the twentieth century, relatively cold surface waters dominated the period from 1890 to 1924 and again from 1947 to 1976. Conversely, warm surface waters prevailed from 1925 to 1946 and again from 1977 to 1998. Presently, colder waters are again dominant in the North Pacific. Notice that the cycle in those ocean temperatures somewhat corresponds to the variations in Midwestern drought. One might ask, "Are droughts responding to solar/lunar cycles or to changes in atmospheric circulation?" Or, "Does the atmospheric circulation of the North Pacific respond to solar/lunar cycles and then in turn impact droughts?" At this time, we simply don't know the answer.

So, unfortunately, with regard to the 1930s Dust Bowl, we are still left with half a mystery. We do know that the part of the misery of those horrifically dry days in the American Midwest was the direct result of the settlers' lack of knowledge of that region and, consequently, their ill-advised farming practices. But the larger question of the ultimate cause of the periodic droughts in the American Midwest still remains elusive to climatologists.

Time: April 14, 2065
Location: Guymon, Oklahoma, North America

This long-lasting dry weather was bad—but not nearly as bad as it might have been, the city manager realized as he glanced up at the framed black-and-white photograph of the infamous 1935 Black Blizzard that had hit Guymon 130 years ago.

Yes, farmers at the local coffee shop this morning were complaining about the continuing drought, but it was more grumbling at the high water-importation fees charged by the GPWRP rather than sheer desperation at maintaining their very livelihoods.

That new water pipeline connecting Guymon with the desalinization plant down at Galveston was a godsend. It was pumping thousands of gallons of good fresh water into the whole bone-dry state of Oklahoma—and, while the cost of that water was undoubtedly high, people weren't migrating to California as the residents of this place did some 130 years ago.

It was just darn amazing that those clever scientists finally figured out what was causing these droughts. At least now they can plan ahead and fund the cost of the "white gold" water pipeline and the Galveston desalinization plant. The crops in the 2040s and 2050s had been good enough to put up as collateral for the federal government to pass the loan for the creation of the GPWRP, the Great Plains Water Reclamation Project.

The city manager glanced out the window where a group of kids was just getting out of school. The kids were running happily through the lush green grass of the city park. *Hard to imagine,* he thought as he glanced back at that stark photo of the 1935 Black Blizzard dust storm, *why people didn't do more long-term planning back then.*

Chapter 17

THE MYSTERY OF THE CRASHING AIRPLANE AND THE A-BOMB

After experiencing ten thousand years of weather, one would think that the people of modern civilization would have achieved a fundamentally complete knowledge of all types of weather. We wouldn't still be discovering any new types of weather, right? And, in any case, our own horrific weapons of mass destruction—nuclear bombs, for instance—couldn't possibly give us any clues about any such undiscovered weather?

Enter our next weather detective, the renowned meteorologist Dr. Ted Fujita.

Time: June 1975
Location: JFK International Airport, New York City, New York, North America

The investigator slowly shook his head as he looked at the still-smoldering remains of the crashed airplane.

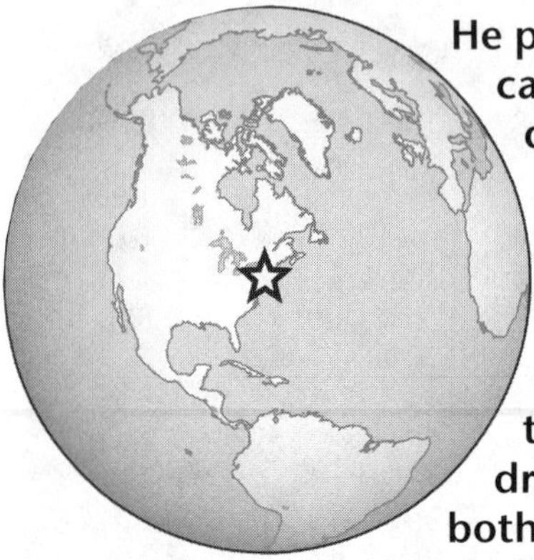

He pulled out a paperback book–sized cassette recorder from his rain-drenched trench coat pocket and flipped it on.

"Continuing notes on the crash of Eastern Airlines 66—a Boeing 727," he spoke into the machine. "Latest word is that the total death toll is now over a hundred people. Pilot and copilot were both killed. We have fourteen survivors—including two flight attendants who were stationed in the back row. That's not surprising given that the plane's largest intact section—about forty-five feet long—is the burned-out portion of the rear fuselage."

He glanced across the now-closed Rockaway Boulevard to the swampy field in which the plane had first crashed. "It appears that the aircraft initially crashed into the field adjoining the airport a full twenty-three hundred feet short of the runway." He tucked the recorder under his arm for a moment, flipped through his small spiral notebook, then pulled the recorder out and once again spoke into the small machine. "Twenty-three hundred feet short of Runway 22-L. Eastern 66 then hit three of the towers supporting the runway approach lights."

He paused, shaking his head a bit. "In perhaps a final heroic attempt to regain control of the aircraft, Eastern 66 lifted and cleared the next three towers—but then proceeded to take out the next four."

"After smashing into the swampy field adjoining the airport, the largest sections of the aircraft—including the tail section—spun across Rockaway Boulevard, ripping through a wire fence before coming to rest."

"Jack." The investigator's colleague walked up through the mucky grasses at the edge of the airport runway. "Sorry to bother you but the media are clam-

oring for an explanation. Do we have any ideas yet to give them?"

The investigator shook his head. "I thought perhaps the plane might have gotten caught in the gust front from that line of thunderstorms in the area at the time of the crash, but the timing just doesn't match up. From the tower, I learned that there was a small Beechcraft Baron that touched down on this very runway just three minutes before Eastern 66 made its final approach. Sure seems to me that a large-scale gust front would have impacted that small Beechcraft a hell of a lot more than this massive Boeing 727!"

His colleague nodded. "But then what caused Flight 66 to literally fall out of the sky some two thousand feet before the runway?"

The investigator pondered the question for a long moment as the drizzle continued to fall around them. He clicked off the cassette recorder and then resolutely looked at his colleague. "I think, Jim, we have to bring in a weather expert on this one. And for this one," he said, glancing over the wreckage of Eastern Flight 66, "I'm pretty sure that we're going to need the best in the business!"

The expert who was brought in to investigate the deadly Eastern Airlines 66 crash was the legendary meteorologist Tetsuya "Ted" Fujita. In conducting that investigation, he made a startling discovery of a completely unknown type of weather, an event that we now call a *microburst.* Before we delve into the science of microbursts, let's first introduce the man who is credited with their discovery. Ted Fujita was born on October 23, 1920, in Kitakyushu City, Japan. He graduated with a bachelor's degree in mechanical engineering from the Meiji College of Technology in 1943 and, by the next year, became an assistant professor of physics at that institute, where he began to study weather. By 1953, he earned his doctoral degree from Tokyo University with an analytical study of typhoons.

But tropical typhoons weren't Fujita's main interest—he was much more intrigued with continental severe storms, the kind that produce tornadoes. When he personally observed a severe Japanese thunderstorm during the summer of 1947 from the wonderful vantage point of a mountain observatory, he proceeded to write a detailed letter about the storm to one of the distinguished meteorologists of the time, Dr. Horace Byers of the University of Chicago. Fujita's letter relayed his speculations about possible conflicting wind patterns inside the storm. Byers was so impressed with Fujita's detailed reasoning that he convinced Fujita to come to the University of Chicago in 1953. It was at that university that Fujita was finally able to devote the rest of his life to the violent weather that so intrigued him. After becoming an associate professor in geophysical sciences in 1962, he quickly reached the rank of full professor by 1965. As part of his work at Chicago, he adroitly directed the Satellite and Mesometeorology Research project and then the Wind Research Laboratory. He is probably best known for the creation of

A photograph of the late Dr. Ted Fujita of the University of Chicago. Photograph courtesy of the University of Chicago Media.

the *F-scale*, which is used to rank tornadoes based on their destruction.[1] In regard to our current mystery, Dr. Fujita was asked in 1976 to investigate the gruesome crash of Eastern Flight 66 at JFK Airport in New York.

What made Fujita the right man to investigate that crash? In short, Ted Fujita was a researcher cut from a different cloth than the typical scientist—he was a master of inductive logic. Inductive reasoning is the process of arriving at a conclusion from a limited set of data rather than from all possible observations. Commonly, if done correctly, inductive reasoning is considered an "intuitive leap." Such reasoning is markedly different from the traditional type of scientific analysis, called deductive reasoning. Deduction—the type of reasoning used by Sherlock Holmes in the A. Conan Doyle stories—involves arriving at a conclusion based totally on previously known facts (called the premises). If the premises are true, the conclusion must be true.

Inductive reasoning is generally considered to be less powerful than deduction because it is much easier to go astray by using induction. The flaw of induction, according to many experts, is that the generalization can break down when new facts materialize because induction involves extrapolation from a limited set of observations. If a person says, based on the available knowledge, that all baseballs are white in color, he or she can be proven wrong if a baseball of a different color is found.

Conversely, deduction involves arriving at the conclusion directly from the observations. Deduction is a step-by-step process of drawing conclusions based on previously known truths. For example, if all fish in a given pond have gills and you catch a fish from that pond, you can deduce that it will have gills. A fundamental problem with deduction is that, if one's initial facts are later proven wrong, the logical chain of reasoning breaks down.

The trouble with using deductive reasoning for an open-ended system such as our atmosphere is that rarely, if ever, are all of the facts known. Consequently, even though most scientists say that they are using deduction—basing their theories on a framework of facts—that framework of facts is, by necessity, limited to only the available set of observations. So, all atmospheric scientists use inductive reasoning to some extent—but some are a lot better at it than others.

Fujita's genius was his well-honed inductive ability—in being able to extract a brilliant fundamental principle from a quite-often limited set of facts. One of his colleagues, Dr. James Wilson at the National Center for Atmospheric Research in Boulder, Colorado, commented upon Fujita's death in 1998 about this aspect of Fujita's scientific personality: "He would theorize how things work, and it often was left to the rest of us to come along and prove his theories."[2] He continued, "There was an insight [Fujita] had, this gut feeling. He often had ideas way before the rest of us could even imagine them."[3]

In the well-ordered world of science, such a renegade attitude led to occasional resentment. Interestingly, many scientists don't like flashes of inspiration or gut feelings. They prefer to undertake their scientific investigations in a straightforward—if rather dull—step-by-step sequence of testing. Wilson noted that Fujita "was a controversial character at times because of the way he did his science. But there was no question that he had insight that very few people had."[4]

Given his preferred methods of investigation, Fujita was not overly fond of computers. This is because inductive reasoning is the complete antithesis of computers, which by their very nature are the ultimate tools for deductive reasoning. The conclusions that computers reach are always inexorably constrained by the initial data fed into them. Computer analysts even have an interesting term that encapsulates the enormous problem inherent in this deductive aspect of computers: GIGO, or Garbage In, Garbage Out. Because Fujita realized that potential flaw in computers could severely limit their usefulness, he seldom used them in his research. A colleague of Fujita's, Chicago meteorologist Duane Stiegler, noted after Fujita's death that "he used to say that the computer doesn't understand these things."[5]

So how did Fujita inductively attack the mystery of Eastern Flight 66's demise?

I was fortunate to have met the brilliant diminutive atmospheric scientist many years ago at a meteorology conference in Phoenix. He told me that some of his first thoughts about the Eastern Flight 66 crash related back to his days in Japan during World War II. He had been one of the first scientists to fly over the desolate city of Nagasaki just weeks after the atomic bombs had been dropped. With that vantage point, Fujita remembered the very particular shock wave—damage pattern of buildings and trees in that city—a radial or starburst pattern emitting out from the central point of the city below where the atomic bomb had exploded. This was in marked contrast to the spiral type of damage that Fujita had observed countless times over the US Great Plains after tornadoes had devastated an area. Trees and buildings felled by tornadoes often demonstrate distinctive spiral patterns that indicate the violent rotation of winds.

But the damage pattern that Fujita saw with the Eastern Flight 66 crash debris was radiating from a central point, sometimes termed *straight-line wind damage*—not spiral! So, Fujita inferred, Eastern Flight 66 was not brought down by flying through a tornado. Were there any other alternatives? Now, perhaps in our post-9/11 world, we might have concluded that the airliner must have been the target of a terrorist bomb—after all, the damage patterns for the Flight 66 debris and the Nagasaki atom bomb were distinctly similar. Luckily for meteorology (and for future airliners), Fujita hadn't eliminated all of the other possible options.

One intriguing possibility that occurred to him was that perhaps a storm cell could sometimes create its own natural "air bomb." What if, he asked, a small storm somehow generated a sudden incredible downward blast of air—perhaps with winds of a hundred miles an hour or more? Wouldn't such a blast create the same kind of straight-line damage pattern that one saw with the Eastern Flight 66 crash?

Previous research had shown that the Great Plains type of storm undergoes a specific life cycle. When a thunderstorm cell—a small individual storm—begins its existence, it consists of air being violently uplifted into the upper atmosphere. As the air rises, it cools and eventually reaches its dew point, the temperature at which condensation occurs, and a cloud forms. As the thunderstorm cell matures, some parts of the thunderstorm cell containing raindrops begin to fall and begin to pull the air down with them. At this stage, the storm has both updrafts and downdrafts. Finally, as the thunderstorm cell begins to dissipate, the overall motion in the storm is downward.

What if, thought Fujita, occasionally the downward winds were incredibly concentrated from a single thunderstorm cell—perhaps when a sudden blast of cold rain forced the air down? We have long known that large-scale downdrafts—sinking air associated with a line of thunderstorms—can occur. For example, in the desert, these massive downdrafts, which crash into the ground and spread outward, have been observed to create massive lens-shaped walls of dust. Such dust walls are commonly called *haboobs* in the Middle East. These gust fronts are formed along the leading edges of large domes of rain-cooled air that result from the merger of cold downdrafts from adjacent thunderstorm cells. At the leading edge of this gust front, there is the dynamic clash between the cool, out-flowing air and the warm thunderstorm inflow that produces the characteristic wind shift, temperature drop, and gusty winds that precede a thunderstorm. These gust fronts were long thought to be the main wind shear threat presented by thunderstorms to aircraft during takeoff or landing.

But, Fujita theorized, what if much smaller individual thunderstorm cells could produce concentrated downdrafts of greater intensity than the more massive gust fronts? The University of Chicago professor conjectured that the crash of Eastern Flight 66 might be due to the impact of a small-scale, jetlike downdraft that Fujita labeled a microburst. Specifically, he defined a microburst as "a small downburst with its outburst, damaging winds extending only 4 kilometers (2.5 miles) or less from a central origin point. In spite of its small horizontal scale, an intense microburst could induce damaging winds as high as 75 m/sec (168 mph)."[6] The key point here is the size of the event. While a downburst involves any sudden downdraft of air, a microburst is a concentrated, small-scale air blast.

In essence, one can picture a microburst as a massive air bomb—a bomb that explodes air down to the ground and then sends it abruptly outward with wind speeds comparable to those of tornadoes. That downward blast of air produces massive wind shear, a rapid change in wind direction or speed. A microburst can produce the severe wind shear with horizontal wind speed changes greater than 15 knots (roughly 17 mph) or vertical wind speed changes greater than 500 feet per minute (around 5 to 6 mph in the vertical direction).

Most aircraft crashes involving microbursts occur when planes are attempting to land. As the aircraft makes its final approach to the runway, the pilots are slowing the plane to an appropriate speed. When the blasting winds of the microburst hit the plane, the pilots would experience a marked reduction in their forward airspeed, caused by ramming into the ferocious headwinds created by the microburst.

I once encountered the same situation as those unfortunate pilots when I was driving a high-profile storm chase van on the ground. We were traveling on a bridge at 60 mph when we encountered the headwinds of a microburst—and literally came to a sudden stop as we smashed into the 100-plus-mph winds of the microburst.

What I did at that time is exactly what a pilot before Fujita's research would have done: I let up on the accelerator and tried to slow down. And that instinctive impulse could be fatal for a pilot. What we know now is when an aircraft encounters a microburst, a pilot inexperienced with these odd weather events usually tries to compensate for the massive headwinds by suddenly decreasing the plane's airspeed. But as the aircraft proceeds to travel through the microburst, abruptly it no longer encounters a headwind but instead is pushed by an incredibly strong tailwind. This causes a sudden decrease in the amount of air flowing across the wings—the critical principle to maintaining flight. Consequently, the sudden loss of air moving across the wings causes the aircraft to literally drop out of the air.

Following Fujita's groundbreaking discovery of these microburst events, we now know that the best way to deal with a microburst in an aircraft is to increase speed as soon as the abrupt drop in airspeed is noticed. This will allow the aircraft to remain in the air when traveling through the tailwind portion of the microburst and also pass through the microburst with less difficulty. One thing that we have discovered since Fujita's time is that commercial jet aircraft are much more vulnerable to a microburst than small planes. In particular, we have learned that a single-engine prop plane can more quickly speed up or maneuver to avoid the consequences of a microburst, while a jet will have a slower response lag that will make it especially vulnerable to the winds of a microburst.

But what causes these microbursts—Fujita's small-scale air bombs—to occur in the first place?

Microbursts are the result of air being rapidly accelerated down from the middle and upper parts of the thunderstorm to the ground. That downflow can occur due to several factors: by air being pulled down by rain or hail, by the increases in air density as the air is cooled by rain, and by the cooling produced with melting ice crystals. These three forces, if strong enough, can create massively intense and sudden downward movements of air.[7]

One aspect of microbursts that we have learned since Fujita's initial work is their strikingly short lifetime. Most microbursts last only five to fifteen minutes. Such limited duration, coupled with their small spatial extent, makes microburst detection and prediction very difficult. But we have made significant progress in detecting microbursts through the use of Doppler radar.

Normal radar bounces microwaves off of storms. By measuring the time it takes the microwaves to return and the strength of those returning microwaves, normal radar can determine the distance and intensity of the storm. Doppler radar extends that principle by measuring the slight change in microwave wavelength that occurs if the target of the microwaves is moving away or toward the radar antenna. By doing so, Doppler radar can establish whether the storm—or the winds within the storm—are moving toward or away from the station. Where normal radar indicates a storm's location, Doppler radar is akin to a weather x-ray machine since by looking inside the storm, it can tell which way the winds within the storm are moving.

Consequently, Doppler radar can detect wind shear—abrupt changes in wind speed and direction—the key characteristic of a microburst. In the years following the Eastern Flight 66 tragedy, airports were fitted with Terminal Doppler Weather Radar (TDWR), which can detect microbursts. And many, many lives have undoubtedly been saved.

In solving the mystery of the crashed plane, Fujita proved that inductive logic can be as powerful and useful in scientific research as its more commonly used counterpart, deductive reasoning. And in using it, Ted Fujita discovered an entirely new form of weather.

Time: June 2010
Location: JFK International Airport, New York City, New York, North America

A flash of lightning lit up the interior of the Boeing 727 flight deck.

The pilot was all business—this was the point when she earned her paycheck.

Damn, she thought, those thunderstorm cells are too close for comfort. This is definitely going to be a bumpy landing.

"Crossways Air Flight 234 now on final approach to R-22," she heard her copilot radio to the tower.

"Take it easy," she muttered to herself as the aircraft wings rocked slightly in the strong winds of the thunderstorm.

"Crossways Air 234! We have a microburst alert signal for your runway! Abort landing."

The pilot immediately throttled up the engines—markedly increasing their speed—and pulled the flight wheel back. For a long moment broken only by the flashes of lightning from the nearby storms, the plane rose gracefully back into the sky at a steep angle. "Radio the tower to give us a safe reroute," she commanded her copilot. He nodded—all business on the flight deck.

As they reached a safe altitude, she keyed the microphone to the back passenger section. "Hello, everybody, this is Captain Poltanos from the flight deck. Sorry for the delay, but we saw a bit of rough weather ahead of us and, to play it safe, we're going to make another pass and try again. We should be safely on the ground in about fifteen minutes."

A loud chorus of groans from the inconvenienced but safe passengers filtered through the cabin door.

Chapter 18

THE MYSTERY OF "ELIMINATING THE IMPOSSIBLE"

Time: The day after "a dark and stormy night"
Location: The Whittingham Manor in England, Europe

The Great Detective carefully studied the faces of the three people watching him nervously in the drawing room of the Victorian mansion. He pulled out a large pipe from his jacket pocket, carefully packed it with tobacco, and lit the pipe with a large wooden match that he causally flicked into the massive stone fireplace. The silence in the room grew to almost deafening levels.

"Today," the Great Detective said after a long puff on the curved Meerschaum pipe, "we will unmask the true murderer of the late Lord Raymond Whittingham." He swiveled on his feet to face the well-endowed young blonde woman seated on the sofa.

"Rita Monroe-Whittingham, as a former exotic dancer and the recently married wife of Lord Whittingham, you stood to gain a fortune by his death."

The woman paled but said nothing under the accusing stare of the investigator.

The keen-eyed detective shifted his attention to a skinny, rather weasel-faced young man leaning against the fireplace.

"And, you, young Alex Whittingham, as the unemployed, debt-ridden son of Lord Whittingham by his first wife, perhaps you thought you could pay off your extravagant gambling losses by killing your father and claiming the inheritance."

The young man frowned but remained silent.

The Great Detective then turned to face the third suspect, a large, portly man seated in an oversized, leather-cushioned chair.

"And, Harold Reamus, as Lord Whittingham's disreputable business partner, you needed to prevent him from uncovering your massive embezzlement of company funds over the last three years."

Glistening beads of sweat appeared on the fat man's forehead.

"But, no, I have ascertained that none of you actually murdered the late Lord Whittingham." The Great Detective paused dramatically, relighting his pipe before continuing.

"No, while each of you had ample motive, none of you had the opportunity or the strength to suffocate him with a pillow as he lay peacefully sleeping during that clashing thunderstorm last night. So, without question, the man who so brutally murdered Lord Whittingham is . . ."

A tall, sober-faced butler entered the room bearing a tray of drinks.

The room was deathly quiet. The Great Detective swung his arm, pointing directly at the somber-faced, muscular manservant.

"Alfred the butler!"

Of course, one of the most overused clichés in mysteries is "the butler did it." But this hackneyed vignette points out a key aspect of nearly every mystery. The classic mystery detective must address all the potential red herrings and remove them from consideration before solving the whodunit and identifying the real criminal. Or, as the Great Detective Sherlock Holmes is alleged to have said, "Eliminate the impossible." Believe it or not, that type of critical inquiry exists not only in mysteries but in the sciences as well. In climatology, the dilemma of eliminating the impossible is most apparent in the need to identify specific biases and other confounding problems existing in the world's weather and climate records.

The brilliant climatologist Reid Bryson of the University of Wisconsin once wrote an article that put this elimination problem in stark perspective. He gave the article the intriguing title ". . . All Other Factors Held Constant."[1] By that, he meant that when a climatologist is attempting to address the impact of, for example, increasing atmospheric carbon dioxide on the world's climate, a critical but often unstated assumption is that everything else that might potentially impact climate has been accounted for and addressed in the weather records.

For example, an inexperienced climatologist might study the temperature record of a city such as New York or Boston and state, "Global warming must be occurring because the temperatures of those cities have steadily increased over the same time that carbon dioxide in the atmosphere has increased." But the problem is that the inexperienced climatologist wouldn't have followed Bryson's rule to hold all other factors constant.

In the case of growing metropolises such as New York or Boston, a critical factor that influences urban temperatures is the increasing amount of asphalt, concrete, and other materials in the city. Such materials effectively absorb the solar heat that falls on the city, much better than trees or grasses do. Consequently, we have found that big urban centers are significantly warmer than they would otherwise be. This warming, called the *urban heat island effect*, is created by the city itself regardless of other factors and exerts a powerful influence on any large city's temperature record. For instance, in the desert metropolitan city of Phoenix, Arizona, my fellow climatologists have discovered that under certain conditions the urban center may be as much as 8°F warmer than the outskirts of the city.[2]

And that's only one confounding influence on a given temperature record. Consequently, if I or any other climatologist wants to identify the global warming signal in a temperature record, potential red herring factors such as the urban heat island effect must be identified and addressed.

The often unreported issue that climatologists must face when establishing the causes of climate change is whether they have truly accounted for all other potential influencing factors, or biases, in the climate records. That is the dilemma faced almost daily by one of my accomplished former students, an exceptional climatologist named Dr. Russell Vose. Vose is one of the top scientists at the National Climatic Data Center, the nation's primary storehouse of historical climate information, located in Asheville, North Carolina. Among his many responsibilities is the task of identifying and eliminating confounding factors in all of the climate records for the United States and for other countries as well.

As noted, one of those red herring biases is the urban heat island effect. For climatologists, an unfortunate aspect of weather measurement is that most of the weather observations around the world have been made in cities. And those cities over the last century have grown exponentially. The temperature measurements and other climate records (precipitation, winds, clouds, and so on) are, unquestionably, influenced by the growth of the cities. Yet we don't want to simply throw out all climate data as irrevocably contaminated by urban growth. So the problem that Vose and other climatologists at the National Climatic Data Center face is how to remove the urban effect from our climate records. How can they safely extract the offending urban influence from the temperature record without removing the other changes, such as global warming, that might influence the record?

One means of combating the urban heat island effect is to use only those locations that haven't been greatly impacted by the growth of cities. For example, Vose and his colleagues identified 1,218 weather stations across the United States that they laboriously determined haven't been contaminated by being in large, growing metropolitan locations. This set of unspoiled weather stations has been called the Historical Climate Network, or HCN, and the latest version is arguably one of the best available records for the study of climate change for the United States.[3]

A temperature sensor, one of the many stations around the world recording weather, outside the Furnace Creek National Park Service Headquarters in Death Valley, California.

But even this record still had potential red herring signals embedded in it that had to be addressed and removed. In particular, Vose and his colleagues identified another—even more obscure—influence that was hiding in most climate records.

Most of the high-quality weather stations across the United States are part of the US Cooperative Observing Network and, as the name implies, are staffed by volunteers. The volunteers themselves set the regular times at which they make their daily observations of the high and low temperatures (using maximum/minimum thermometers, for instance). Vose and his colleagues determined that this simple situation can introduce bias into the temperature record.[4] Because most individuals don't make their measurements at midnight, their daily observation period—the time from one day's readings to the next for which the high and low temperature measurements correspond—doesn't conform to the standard calendar day's midnight-to-midnight time frame.

If a weather volunteer makes his or her readings at 8 AM each day, then the high and low temperatures are effectively the high and low values recorded between 8 AM of the previous day and the current 8 AM measurements—but they are listed as the official values for the current day. However, suppose another observer likes to take his readings at 5 PM. For that individual, his observation day of high and low temperatures is a record for the times between 5 PM of the day before and the current 5 PM readings. And as with the morning person, these are listed as his official values for the current day. Both observers log their values for the same calendar day but there is the total of an astonishing nine hours' difference between the true observation day of one and the true observation day of the other. Perhaps surprising to realize, that simple difference in when the observation is made can lead to major spurious climate change data introduced into the long-term climate record.

Vose and his colleagues at the National Climatic Data Center determined that as much as 1.5°C (almost 3°F) of error can be introduced into the monthly mean temperature by simply having differences in the time of observation. This is because, in a given April, for example, an observer who is recording in the morning will have perhaps as much as nine hours more of his observations from March incorporated into the April monthly record (since on April 1 the observer making his observations at 8 AM will in essence be using hours of measurement from the previous day in March). By the end of the month, however, the afternoon observer will have perhaps as many as nine more hours of April 30 values than his morning recording counterpart. This means the temperatures recorded by the person who makes her observations in the afternoon will average slightly warmer than the person who makes her observations in the morning over the long term.

Seems trivial?

Not when you consider that we are discussing global climate changes on the order of fractions of a degree over decades. It turns out that the relatively small time of observation bias is big enough to potentially obscure our true findings of climate change. Therefore, identifying, quantifying—and then removing—those biases from our databases of climate is absolutely critical if we truly want to understand at what rate our climate is changing.

But perhaps you are thinking, "Just how many of those potential biases can really exist in our climate record?" Undoubtedly, by this time, we have found them all, right?

We keep finding them.

For example, one bias existing in temperature records that my colleagues (including the esteemed Russ Vose) and I only recently uncovered involved our fundamental calendar system itself.[5] Our current calendar system, the Gregorian calendar, was designed to keep the spring equinox (one of two days each year when every location on the earth has twelve hours of day and twelve hours of night) on or close to March 21, so that the date of Easter (observed on the first Sunday after a full moon that falls on or after the spring equinox) remains correct with respect to the spring equinox in the Northern Hemisphere. The Gregorian calendar decreed by Pope Gregory XIII was developed as a reform of the Julian calendar. A key feature of the Gregorian (and of the earlier Julian) calendar system is the incorporation of periodic "leap years," that is, years containing an extra day. Their purpose is to keep the calendar year synchronized with the astronomical calendar. In particular, in the Gregorian calendar, February periodically contains twenty-nine days instead of the usual twenty-eight.

Similar to the time of observation bias, the incorporation of an extra day means that the actual temperatures we record for a given day aren't really for that day (seasonally) every fourth year. Conceptually, the climatological effect of adding an extra day in February is to shift the average of the monthly temperatures for February through June into warmer periods of spring and summer. The effect is reversed, however, for the months of July to December, where the addition of an extra day in February effectively shifts the average of those monthly temperatures into the colder periods of autumn and winter. When we tested the impact of this bias on the world's best and most used climate change data sets (the ones that we hear reported by the media with every global warming study), we found that spring months in leap years are slightly warmer than the spring months of common years (nonleap years) and that autumns in leap years are slightly colder than their common year counterparts.

So the mystery continues. As the Great Detective reportedly said, only after eliminating the impossible, the remaining—no matter how implausible—must be the truth. For climate scientists, one of our biggest concerns with today's climate change research is simply to discover all of the impossibles . . . and to eliminate them.

Time: The day after "a dark and stormy night"
Location: The Whittingham Manor in England, Europe

Alfred the butler gazed dispassionately at the accusing finger of the Great Detective.

"But, esteemed sir, I could not possibly have committed the brutal murder of my master, the late Lord Whittingham," the butler protested. "Before that horrible storm broke last night, the master had released me for the evening and so I was in residence at the local village pub, the White Hart. The storm's rain, thunder, and lightning made the night not fit for man or beast so I stayed at the Hart. Several of the local men of the shire can vouch for me."

The Great Detective frowned.

This complicated the mystery greatly.

"I know who killed Lord Whittingham!" A new voice—that of a mysterious Climatologist—broke the silence of the room as the bespectacled, red-haired scientist strode confidently into the room and gazed at the assembled people.

"Yes," the Climatologist continued. "This is a perplexing mystery that can only be solved after all of the conflicting factors have been addressed. And there is one thing that our Great Detective has overlooked in his investigation!"

Stunned silence filled the room.

"At the moment of Lord Raymond's demise, Whittingham Manor was at the exact center of a massive lightning storm. So I have examined the lightning strike record of that storm and correlated it to the precise moment of Lord Whittingham's death."

The Climatologist paused dramatically.

"It only appears that Lord Whittingham died of suffocation. I discovered his lordship actually suffered a massive heart attack after being struck by lightning that was conducted into his room through the fireplace. The evidence clearly indicates that it was the hand of nature—not the hand of man—that struck down the unfortunate Lord Whittingham!"

Chapter 19

THE MYSTERY OF THE DEVIL IN THE DEEP BLUE SEA

As we saw with the Mystery of the Pacific Hot Tub, scientists are discovering that some of the most important controls on the world's atmosphere are, perhaps somewhat surprisingly, the world's oceans. For example, the discovery of the interrelationship between atmospheric winds and oceanic currents in the central Pacific Ocean, called the El Niño–Southern Oscillation (ENSO) phenomenon, has been one of the great advances in climatology. Through our knowledge of ENSO, we now have substantially better seasonal weather forecasts than we did just a couple of decades ago.

Other aspects of the oceans may play an even more fundamental role in determining our climate. One oceanic feature, called the *global thermohaline circulation*, has only recently been identified as having a critical influence on climate change. In fact, that ocean feature may be so controlling of our weather that it is perhaps the crucial element to determining whether we will stay warm or have to endure another ice age. Parts of the oceans' potentially devilish thermohaline circulation were first discovered more than two hundred years ago.

Time: Fall, 1768
Location: London, England, Europe
(This and the closing vignette of this chapter are loosely based on a letter from Benjamin Franklin to Mr. Alphonsus le Roy that appeared in the *Transactions of the American Philosophical Society* in 1786.)[1]

The portly bespectacled gentleman of a respectable sixty years of age or so set his tankard of ale down with a satisfied sigh of contentment.

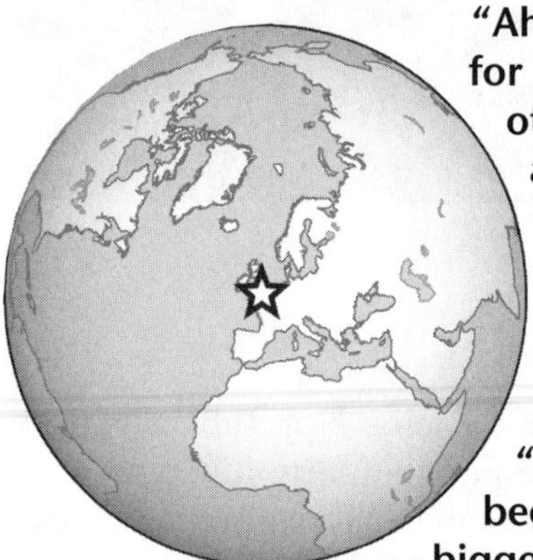

"Ah, my sincere appreciation to you for agreeing to meet with me, Timothy," he said to his companion across the table, a grizzled fellow American of rather small stature.

"Indeed, cousin Benjamin," the sailing master nodded, as he also set his tankard onto the rough timbers of one of the pub's tables. "I must be thankin' you for the fine beer but I thinks that you have a bigger purpose for callin' on me."

The generously proportioned older man smiled. "I would hope, Timothy, that I always have the time to savor a good pint of ale with a dear friend and relative, and a fellow American. But you are correct; I have been asked by the British Lords of the Treasury to contribute my scientific knowledge to a matter that perplexes them greatly."

The whaling captain nodded. "Anything I be able to do to help you, Benjamin, you need only ask. Rest assured that we back in the colonies know what a good job the great Mr. Franklin be doing to help us on this side of the sea."

"My thanks, good captain," the scholarly gentleman graciously accepted the compliment. "The crux

of the matter appears to be the incredible speed of your—and your fellow Rhode Islanders'—ships. As you probably know, I am the postmaster general for the American colonies. One thing that I have noticed is that it appears to take the British mail packets much longer to sail the ocean to America—sometimes by as much as a fortnight—than it does for you traders from Providence. I would very much like to know the secret behind the prodigious speed of the ships of my cousin, the great Captain Timothy Folger."

The whaler crinkled his weatherworn face into a crafty smile.

"Ah, you be wonderin' why we always be beatin' 'em across the great sea, do you?"

The scientist-statesman nodded, carefully keeping his face impassive as he peered at the small captain through his bifocal glasses.

The American sailing master leaned over the small table. "Be it you, cousin Benjamin—who be doin' the askin', I be tellin' you."

The portly gentleman waited patiently.

"It be the waters, cousin," the whaling captain whispered, imparting the great revelation. "We Rhode Islanders know that there be a strange streamlike current in the great sea between the colonies and these British Isles. It be a stream or river of fast-moving waters that runs along the coast of the colonies and then far north across the northern seas. We be well acquainted with that stream because of our pursuit of whales, that keep near the sides of it. If you be wise to those waters, Benjamin, you sail north and reach the Isles here long before those dawdling Falmouth boats."

The little sea captain smiled, "I once spied a British packet hauling westward into the stream and advised 'em to cross it and be free of it, but they be too wise to be counseled by a simple American fisherman. They continued to plod against the stream and we beat 'em to port by a good week."

The portly gentleman nodded. "The British do have

difficulties admitting their wrongs. But I am intrigued. A river in the sea, eh? It is a pity no notice has been taken of this stream upon the sea charts."

We have known the existence of the Gulf Stream—that strange surface current of warm waters along the East Coast of North America—for the relatively short time of only the last few hundred years or so. Franklin and his whaling cousin were among the first to determine that the Gulf Stream is first noticeable in the Caribbean Sea and then flows northward along the East Coast.[2] More important, we have discovered only in the last few decades that the Gulf Stream is part of a much larger global circulation, called the *global thermohaline circulation*, that redistributes ocean water around the entire world. The impact of that circulation on the world's climate is the focus of this mystery.

The global thermohaline circulation involves the concept that large-scale movements of ocean waters are created by the movements of heat and salinity inherent in those waters. The amount of salt and the temperature of the water both create changes in its density. Fresh water is less dense than salty water. In a similar fashion, warm water is less dense than cold water.

Since the waters of the Gulf Stream are warm and therefore less dense than the colder waters underneath, it exists as a surface current. But as those waters are pushed into the North Atlantic, they begin to cool. Other waters from the St. Lawrence Seaway—less salty waters—flow into the region and push down the saltier and increasingly colder North Atlantic water. As those waters are pushed down, they begin a long journey back south along the bottom of the world oceans into the South Atlantic, along the coast of Antarctica, and eventually into the Pacific. Once in the eastern North Pacific, those deep ocean waters have warmed and lost enough salt to begin to rise up toward the surface. At this point in their journey, those waters resume a southerly migration again—this time at the surface—through the Indian Ocean, back to the Atlantic, and eventually become, once again, the waters of the Gulf Stream.

This massive recycling of the world's ocean waters has been colorfully labeled by one of its discoverers, a famous oceanographer named Dr. Wallace Broecker of the Lamont-Doherty Earth Observatory, as the *great conveyor belt*.[3] Broecker demonstrated that the waters of the world's oceans continually replenish and deplete themselves of salt and heat as they work their way across the globe.

While such a discovery is undoubtedly of critical interest to oceanographers and others who are interested in the world's waters, how does such a basic recycling of ocean waters via salinity and thermal changes impact the world's climate?

The answer to our mystery involves research by an old teammate of mine, an exceptional glaciologist named Dr. Richard Alley of Pennsylvania State University. Glaciologist? How can a glaciologist—a scientist who studies ice sheets and glaciers—don the garb of a climate detective and explain how the thermohaline circulation could impact the climate?

Alley's work focuses on how ice sheets grow and decay. He has worked on the glaciers in Greenland, on the great ice sheets of Antarctica, and numerous places between those two. I first met him in Antarctica many years ago when I was working on that remote icy continent to drill a series of small holes into a fast-moving glacier

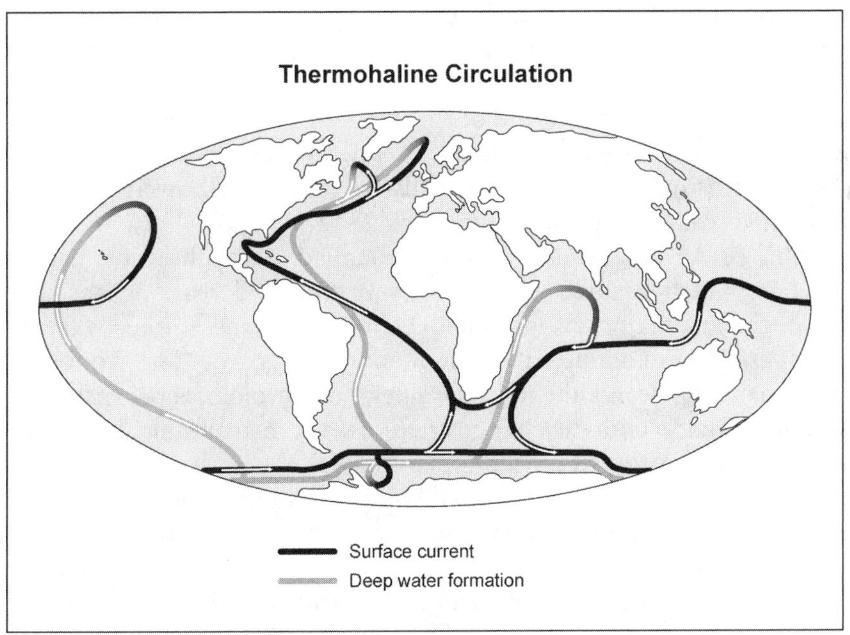

The great conveyor belt of heat and salt known as the global thermohaline circulation. Graphic by Becky Eden.

near the South Pole. Our hope was to eventually achieve a better understanding of why that particular glacier was moving so fast.

But a substantial part of Alley's work—at least with regard to our current mystery—has involved the glaciers of the Northern Hemisphere and in particular what happened to a couple of massive continental ice sheets that are no longer with us, the Laurentide Ice Sheet of North America and the Scandinavian Ice Sheet of northern Europe. These two ice sheets melted away at the end of the last great ice age some ten thousand to twelve thousand years ago. A key question of climate science has been to determine exactly what happened to Earth's climate when those ice sheets disappeared. Alley has stated that the answer to that straightforward question has profound implications for our future existence on Earth.

Alley arrived at his somewhat alarming findings in part from research by Hartmut Heinrich who, in the late 1980s, saw evidence of periodic high quantities of continental rock in layers of deep-sea sediment cores of the North Atlantic Ocean.[4] Heinrich theorized that such rocky debris was carried far out into the ocean by the massive discharge of glacial continental ice, sometimes nicknamed "ice armadas." These ice armadas occurred when the Laurentide Ice Sheet of North American and the Scandinavian Ice Sheet in Europe melted away. The time period of these extensive ice cascades into the Atlantic, called "Heinrich events," tended to coincide with rapid, short-term (I'm speaking geologically—less than a thousand years) cooling across large parts of the world.

One of the best studied of these so-called Heinrich events is the last one, which occurred around eleven thousand years ago at the end of the last major ice age. That event is called the *Younger Dryas cold period*, and it lasted about a thousand years. The Younger Dryas has long been a thorn in the side of climatologists because one of our primary climate change theories, the astronomical orbital change theory, simply doesn't explain it.

But the concept of the thermohaline circulation of the world's oceans provides a new way to address the complexities of the Younger Dryas. Broecker, Alley, and others reasoned that, at the end of the last ice age, as the large Laurentide Ice Sheet of North America was in its final throes of melting, it released a gigantic discharge of freshwater icebergs into the North Atlantic. The particular location of this discharge—the North Atlantic Ocean—was critical because it coincided with the region where the thermohaline circulation's warm surface waters undergo the transition of becoming cold deep-water flow. If that part of the Atlantic Ocean suddenly was flooded with vast amounts of fresh and less dense waters from the melting icebergs, then the entire global thermohaline circulation would simply break down.

Alley noted that the Younger Dryas cold event has not been matched (in size, duration, or rapidity of onset) by any other climate change episode since then. He drew the rather unsettling conclusion that what we call "normal" climate—the climate of the last few thousand years—"has been more stable in any or most regions than is typical of the climate system."[5]

This led him to investigate other periodic cold pulses that have occurred in our climatic past. These cold pulses, termed *Dansgaard-Oeschger events*, started as very short periods—here on the order of decades—of very rapid warming in the Northern Hemisphere, particularly in the North Atlantic Ocean. This was followed by gradual but substantial cooling over the next few centuries. Alley was particularly intrigued with the apparent quickness of these D-O events.[6]

As you can imagine with today's interest in global warming, the idea of marked warming of the North Atlantic leading to the formation of, at the very least, a "little ice age," was something that captured the interest of the media and the general public. Indeed, the special effects blockbuster movie *The Day after Tomorrow* used that concept (but few of the specifics) as the foundation of its core disaster.

The trouble with applying the D-O events to modern times is that we haven't completely solved the mystery. We don't know the fundamental cause of all D-O events. Most likely they are linked to changes in the North Atlantic Ocean circulation, perhaps, as was suggested with the Younger Dryas event, triggered by a sudden influx of fresh water via icebergs. Other possible causes have also been postulated such as involving changes in solar radiation (perhaps, for instance, involving solar storms), or even concerning some still unknown variations in deep ocean currents.

But no matter what their eventual cause, as Alley noted, D-O events like the Younger Dryas have created "climate shifts up to half as large as the entire difference between ice age and modern conditions" and have taken place "over hemispheric or broader regions in mere years to decades." Why is that important? We like to think of the world's climate as having been largely stable throughout time. That comforting thought has led many people to believe—falsely according to Alley and others—that only recently has our climate become more variable. Unfortunately, through the research of Broecker, Alley, and others, we are beginning to understand that Mother Nature has long been an exceptionally fickle mistress—and will likely remain so in the future.

Time: October 1789
Location: Philadelphia, Pennsylvania, North America

The elderly scholar, politician, and—in the minds of a few on the Continent—general troublemaker, sat back in his chair with a deep sigh.

I am getting old—in some ways, eighty years has been too long of a time to traipse across this marvelous world, **the man thought, as he dipped his quill back into the blue-glass inkwell. He coughed a bit raspy. The "stone" in his lungs seemed to be aching a bit more today.**

Indeed, **the Philadelphia gentleman pondered as his breathing became normal again,** ***I may never have another occasion to write on this fascinating subject of the Gulf Stream so I had best write it all down since Mr. le Roy was so kind as to request it of me after our recent talk.***

He thought of his good cousin, Captain Timothy Folger, that extraordinary Rhode Island whaler. After all of these years, it still amused the old man to remember how Folger had smilingly told him how the British had obdurately ignored the advice of a "simple American fisherman" and stubbornly continued to plod against that powerful stream in the ocean—allowing those simple American fishermen to beat the British ships time and time again across the great ocean.

Oh, yes, **the elderly scientist through to himself.** ***One oddity about the Gulf Stream that I simply must divulge to my colleagues is that odd aspect of its temperature.*** **He dipped his quill into the inkwell again. "The Gulf Stream," the former politician wrote, "is always warmer than the sea on each side of it."** ***I must***

remember to append my own sea temperature measurements to this letter, he told himself. Interesting about that temperature difference, though.

He dipped the quill into the ink again, "It would appear from my observations that the thermometer may be a useful instrument to a navigator, as the currents from southern seas into northern are found warmer. And it is not to be wondered that so vast a body of deep warm water, several leagues wide, coming from between the tropics and issuing out of the gulf into the northern seas, should retain its warmth longer than the twenty or thirty days required to its passing the banks of Newfoundland."

Definitely the fogs of the Grand Banks must be caused by the presence of that great Gulf Stream. The weather of that region is profoundly influenced by the stream's consistent hot waters. *Consistent hot waters,* he thought, *that is a key phrase. It might be interesting someday to speculate upon what could happen to the region's weather if the Gulf Stream were ever to cease flowing.*

Chapter 20
THE MYSTERY OF CLIMATE, COCAINE, AND THE STORYTELLER

Without question, drug trafficking has become one of the primary plot elements in today's mystery genre. Because of their illicit nature and big money payouts, narcotics operations are textbook breeding grounds for all manner of mystery-generating crimes, from bribery to smuggling to even murder. But, as the previous mysteries of this book have shown, climatology is one of the most interdisciplinary of all of the world's sciences. Weather and climate literally touch every aspect of our society. Strangely enough, climatology even influences clandestine activities such as illicit cocaine production in a very remote part of the world.

Time: Sometime in the late 1960s
Location: A secluded farm on the edge of the Altiplano, Bolivia

Manuel sighed as he gazed out at the coca plants laid out on the drying tarps near the harvested fields of his farm.

"Si, Senor, it is a very hard life. I am very sad. It will be very hard to make enough money on this crop to feed my wife and children this year."

The young, somewhat brash American geographer smiled sympathetically. He had the rare gift of being able to talk with almost anybody, sometimes when he didn't know the language! This was a great place and time to put that skill to work, he realized as he took mental notes of the environmental conditions in this remote place.

"But your harvest is very good, Manuel," the American countered with a somewhat puzzled tone. "There have been good rains over the past month. I see many harvested coca plants here." He gestured to the many plants carefully laid out on the drying tarps.

"Ah, that is the very problem, Senor," the farmer replied very disconsolately as a clap of thunder heralded the start of a heavy rain shower. "That is the problem."

On the shelves of my rather messy office, I keep a large set of three-ring binders with detailed notes on my past investigations (yes, I really am that old of a professor). Each binder is carefully labeled as to the subject of its contents. Most have rather mundane titles such as "Radiosonde Data," "Astronomical Modeling," and so on. But one binder in particular draws greater attention from students and other visitors to my office than any other. That binder is labeled "Climate and Cocaine," and it demonstrates more than any other investigation that I have ever undertaken why the academic field of geography is an important science.

When I started college as an undergraduate at the University of Nebraska, I was an electrical engineering student. Why? All of my close friends were engineering students and, while we weren't completely aware of what engineers actually did, we did appreciate the fact that engineers earned very good salaries. But one day after working through my umpteenth circuit diagram, I realized that I simply didn't like doing that kind of work. Nothing against electrical engineering (many of my closest friends are electrical engineers and, after all of these years, some are prominent ones involved with the construction of deep-space probes and the like), but I wanted something else. So I asked my knowledgeable brother who was then an admissions counselor at Nebraska (and is now that institution's esteemed dean of admissions), "What department of the university works with weather?" After a moment's searching, he informed me that it was the department of geography.

Geography, I thought with some misgivings. Isn't that just learning places' names and capitals?

Well, I quickly discovered that geography is, literally, the mother of most disciplines, the basis of everything from anthropology to zoology. As history is to studying time and events, geography is to studying place. Everything on this planet has a geographical component. Consequently, weather and climate, obviously, are geographical.

But why not have climatology in, for example, a department of atmospheric sciences? Why place climate science in what some people might term the rather ordinary field of geography?

Part of the answer has to do with the newness of climatology. As I mentioned earlier, the science of climatology is only a few decades old. That means until it gets enough credibility (and students), the academic study of climatology will be usually put into other existing disciplines of a university. At the time I was at the University of Nebraska, that discipline was geography. Indeed, once I graduated with my doctorate and eventually came to Arizona State University, I was also accepted into that institution's Department of Geography (now called the School of Geographical Sciences). But a part of the answer of why climate studies are often in geography has to do with the idea that climatology involves the detailed study of one aspect of the world. And the discipline whose charge it is to study the entire world is clearly the field of geography. Consequently, it is reasonable to put climatology into geography. But that has led to perception problems by those in some of the other academic disciplines.

Unfortunately, many scientists and engineers have tended to look down at the field of geography, often considering it to be a softer science than physics or chemistry. Part of that opinion involves the fact that geography is a two-faced discipline. While a goodly number of us study physical geography, the variations of natural stuff on this planet, from plants and animals to rocks and landforms to climate, a sizable group of geographers are deeply concerned with human geography, the study of where, how, and why people have settled, developed, traded, and moved around on this planet. It is a regrettable fact that engineers and many "hard" scientists sometimes don't think highly of human-based (or social) sciences. One of my favorite authors, Robert A. Heinlein, an ardent supporter of engineering, summed up that attitude by proclaiming, "If it can't be expressed in figures, it is not science; it is opinion."[1] Having grown up with engineering friends and even starting out as an engineer myself, I must admit I had some of that attitude.

That is until I arrived at Arizona State University and met a very interesting geography professor by the name of Ray Henkel. Henkel is an unabashed human geographer. He is captivated by all aspects of the human condition and, because of that, he has traveled extensively and explored some of the most interesting and unusual places around the world.

Now, a notable aspect of academics is that professors tend to interact mainly with the colleagues who are located near them. In most university departments, that means scientists are likely to discuss their work with colleagues who are probably doing similar work to their own. But in a department of geography with such a dichotomy between human and physical elements of the discipline, it means that frequently a professor encounters someone whose work seemingly has little or no relation to her own. And that was the case with Dr. Ray Henkel and me.

One day in the early 1990s I was discussing with him exactly what a climatologist does, telling him that we try to relate climate changes to various causes, such as El Niño–Southern Oscillation or volcanoes. After patiently listening to me yammer on, he suddenly asked me, "Okay, but aren't you also concerned with climate's impact on society as well?" I replied that of course I was—but did he have a specific idea in mind?

He proceeded to tell me a story—and, as all of his colleagues and many past students will tell you, Ray Henkel loves to tell stories.

It seems that sometime in his early career many decades ago,

Ray had been working for some government agency (specifics aren't always given in a Henkel story) in Bolivia. During that time, he had a chance to work with a lot of people of many different types and backgrounds. Some of those people were young farmers working on the fertile soils of the area near the Altiplano, the high plateau that composes most of the country of Bolivia. Others were new government officials who were just working their way up through the bureaucratic ranks. As he was wandering through the countryside, he noticed a very particular aspect in the harvesting of one of the most important crops of the region, coca.

Coca, at that time before the boom in drug trafficking, was consumed locally often as a sedative used to counter the effects of such things as altitude sickness. Locals told Henkel that the quality of their harvest of coca was related to whether or not it rained.

Knowing absolutely nothing about the harvesting of coca plants, I interrupted Ray's story and said, "Oh, the Bolivian coca harvest is better in years with lots of rain?"

Ray flashed his engaging smile. "No, actually, it is just the opposite."

He proceeded to tell me that one of the critical steps in the production of cocaine from coca plants involves the drying of the coca leaves after harvesting. The coca leaves are spread out in the open air to dry—similar in nature to the curing of tobacco leaves. Since those regions of remote Bolivia generally don't have the materials or funds to build large drying sheds, the leaves are often left to dry in the open air.

"And, according to the locals there in Bolivia, if it rains during the time that the leaves are drying," Ray commented, "then the quality of the cocaine produced dramatically drops."

"So frequent rains actually lead to lower cocaine production?" I asked.

"That would be my guess."

"Say, Ray," I asked, with the glimmerings of an idea beginning to form in my mind. "You wouldn't happen to have access to any actual cocaine production values, would you?"

The intriguing answer to that question was yes. It appeared that some of Ray's young farmer acquaintances who he had met in those early times had over the intervening years prospered as the price of cocaine on the world market had risen. Their farming operations had expanded—and they and others had kept rather detailed records of their past harvests. Sure, Ray said, he could get some raw data involving cocaine production for past years in Bolivia.

"And, Ray, you wouldn't happen to have any weather data from that region, would you?"

Again, the answer was yes. It turned out that a number of Ray's young government friends in his early visits to Bolivia had also used the years profitably to advance through the levels of Bolivian bureaucracy. A telephone call or two and we had a very nice set of weather records for locations captivatingly close to regions where coca was grown.

So we had a set of data involving annual coca production in Bolivia and a set of data of the associated rainfall conditions in Bolivia. Together they composed a tailor-made study for accessing if the effect that Ray Henkel had noticed was correct: that frequent rain lowered coca production.

We pulled a talented graduate student, Mike McGlade (now a professor at Western Oregon University), into the study and began to crunch numbers. A battery of basic statistical analyses confirmed our hypothesis; in some cases, an incredible 64 percent of the annual variability in coca production was accounted for simply by variations in the number of rain days for that area. And, as Ray had surmised from his contacts in Bolivia, the relationship was inverse; that is, more rain days actually meant lower coca production. The reason, we speculated, wasn't climatic as much as agricultural; coca needed two to three days of drying time for a good harvest. The more wet days, the lower the coca harvest (and the lower the production of cocaine). Ray's simple ambling through the countryside of Bolivia had led us to an important social/climate interaction.

This mystery points out a couple of significant points. First, it demonstrates the incredible power of simple, diligent observation. Second, it shows the power of thinking outside of the box, in this case linking very disparate subjects like cocaine and climate. As this approach is often used in geography, it can be a very useful skill when done properly. Our climate detective for this mystery, Ray Henkel, has those two qualities in spades. While certainly not an orthodox scientist according to Heinlein's "express it in figures" definition, Henkel certainly is a scientist by another criterion of Heinlein's: "The difference between science and the fuzzy subjects is that science requires reasoning, while those other subjects merely require scholarship."[2] Although he does prefer storytelling to crunching numbers, Ray Henkel—a superb observer who reasons from his observations—is definitely a great scientist.

Perhaps you might be asking at this point, "What happened to that research? Was it useful?" For us, academically, it certainly was. We published our climate and cocaine study in one of the world's most esteemed science journals, *Nature*—a technical journal highly respected by many traditional "express it in figures" scientists.[3]

But, given the high-profile character of that journal, you might be surprised to learn that there has been very little follow-up—at least in terms of published academic papers—to our study that showed more rain lowers cocaine production. Perhaps that observation just points out that not all the tangible results of a science study are necessarily found buried in the pages of technical science journals.

Time: September 2010
Location: Flying high over a secluded farm on the edge of the Altiplano, Bolivia

"Conditions nominal for drop," the weather ops officer radioed to the pilot. The modified C-130 transport displayed no obvious marking indicating its organization or country of origin.

"Ready the sprayers for payload release," the pilot commanded her flight deck officer. "On my mark . . ."

"Acknowledged," the flight deck officer relayed back. "Emitters are operational. Waiting for your mark."

The flight deck officer stood patiently by the release switch for the silver iodine emitters.

"Now," the pilot ordered.

The flight deck officer hit the release switch. The large cargo plane transformed into a mammoth cropduster as it spewed out large amounts of silver iodine into the puffy cumulus clouds of the Altiplano borderlands.

After a few minutes, the pilot turned with a grin to her copilot and navigator.

"All right, folks! We've got a good run under way! Those clouds are growing nicely."

"And, Captain, radar is now indicating budding rain showers over the area," the copilot confirmed with a smile himself.

The pilot's grin grew. "Looks like we're going to have a good response to our cloud seeding this time, Juan. Now, if those egghead scientists are right, we may have just rained out a significant chunk of the Bolivian drug trade for this year!"

Chapter 21
THE MYSTERY OF THE FUTURE

I have spent much of my professional life unlocking the secrets of our world's climatic past. One rather nice aspect of having a historical perspective of the world is that delving into the past rarely leads to unfortunate consequences, such as ridicule, or, in extreme cases, death. Unfortunately, history and myth contain many instances when the opposite holds true. Folklore and historical chronicles show us that when a person proclaims the ability to predict the future, many people will take serious exception.

Time: A time deep in the mists of legend and myth
Location: The palace of the king of Troy, Asia

Cassandra, daughter of King Priam, shook her head with a sly smile as she stretched sensuously on the luxurious bed in the palace. "I'm sorry, my divine love, but I now have exactly what I want—and that doesn't include you."

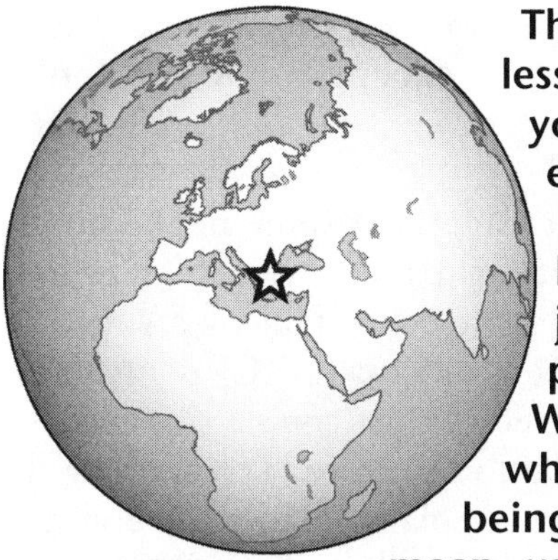

The incredibly handsome, beardless young man gazed upon the young princess with widened eyes. "You deny me?!"

"Listen, Apollo," the flaxen-headed vixen laughed, "You've just given me the ultimate divine power of prophecy, of foresight. With such an incredible ability why do I need to consent to merely being your passing love interest? I mean, you may be a god but now with the ability to predict the future, I have the power of a goddess! I don't need you anymore!"

For a long moment, the golden deity stared at the smug princess lying on the bed. Then abruptly, he laughed.

"You believe my gift of prescience gives you the omniscient power of a goddess, my beautiful but foolish little mortal?" the god asked with amusement. "I do not think you fully comprehend the . . . shall we say, *complete* ramifications of my wonderful present to you."

The young man transformed—growing in stature until his head reached the high ceiling. His youthful visage changed into the awe-inspiring image of a mature—and very angry—deity.

"Know, foolish and insolent tease, that indeed you now and forevermore shall have the power of foresight." The god's intense fiery gaze on the young girl altered into one of pure hatred. "But, coupled with that power is this corresponding curse from the Golden God whom you have so unwisely rejected: From this day forth, tho' you will accurately perceive all of the strands of the Weave of Life, none—not your kinfolk, your friends, nor even your enemies—will ever believe a single word—nay, even a solitary iota—of your precognitive divinations. Be doomed to all from this moment on to be the charlatan prophetess, to be the seer of false foretellings!"

With a flash of brilliant golden light, the bedchamber was abruptly empty except for a suddenly sobbing young princess.

A few years ago I was asked to create what some might say is absolute folly, or what others might term the "ultimate weather forecast." I was hired as an expert climate consultant by a subcontractor to the Nuclear Regulatory Commission (NRC) to make a very long-term climate prediction. Over the last couple of decades, the NRC has built a mammoth installation in Nevada to contain all of the nation's nuclear waste from power plants across the county. This waste repository is housed inside the hollowed-out interior of Yucca Mountain, a facility somewhat near the notorious Area 51 of UFO fame. But in order for the Yucca Mountain facility to be designed properly (so there are no leakages over the lifetime of the storage system, for instance), the future climate to cover the nuclear materials' radioactive half-lives needed to be known. Consequently, regardless as to the philosophical and social problems involving the usefulness of such a facility, four truly brilliant climate colleagues (Henry Diaz from Colorado, Peter Robinson from North Carolina, Tom Wigley from Great Britain, and Cort Willmott from Delaware) and I were given the task to create a "ten-thousand-year" forecast for the future weather of Yucca Mountain, Nevada.

A daunting task?

Surprisingly, a ten-thousand-year weather forecast is fairly straightforward. Such a long-term forecast is possible primarily through the work of one man, a Serbian astronomer named Multin Milankovitch. In 1941, Milankovitch published a detailed set of mathematical calculations that computed the amount of solar radiation received over the world, extending back through the last hundred thousand years.[1] These calculations were based on the very long-term, relatively small changes in Earth's orbit. The orbital variations of our planet influence the amount of solar radiation it receives and, therefore, they are critical to the long-term climate of our planet.

The orbital changes responsible for changing the amount of solar radiation received include the *eccentricity*, or shape of the Earth's orbit; the *obliquity*, or tilt changes of Earth on its axis; and the *precession of the longitude of perihelion in relation to the moving vernal equinox*. This last mechanism is a bit more complicated, so let's discuss it last.

First of all, eccentricity refers to the variations in the shape of Earth's orbit. At the present time, Earth's orbit around the sun is not a perfect circle; that means there are times when we are slightly closer to the sun and times when we are more distant. The day of the year when we are closest to the sun (called by astronomers *perihelion*) is currently July 5, while the day when we are most distant from the sun (termed *aphelion*) is currently January 4. Consequently, the orbit of Earth is a bit more oblong-shaped rather than circular. However, over a long period of time, Earth's orbit gradually shifts from being almost circular to being markedly oblong shaped and back again to being circular. The time required for Earth's orbit to undergo such a periodic transition is roughly a hundred thousand years.

Obliquity changes involve variations in Earth's axial tilt relative to its orbital plane. Earth's orbital plane is the imaginary surface that intersects the centers of Earth and the sun. Planetary axial tilt is responsible for Earth's seasons. The tilt of the planet is measured by the difference between Earth's axis (the line connecting the North and South poles) and the orbital plane. Currently, the tilt of the planet is 66.5° from the orbital plane (or as cited more frequently by scientists, 23.5° from the vertical to the orbital plane). Earth's tilt establishes the latitudinal location receiving the direct (perpendicular) rays of the sun, that is, a point whose solar elevation angle is 90°. That point is called the *subsolar point* and it migrates over the course of the year from the Tropic of Capricorn at 23.5°S during the Northern Hemisphere's winter solstice. It then moves across the equator to the Tropic of Cancer at 23.5°N during the Northern Hemisphere's summer solstice and again back to the Southern Hemisphere by winter. The greater the elevation angle at any given location, the more direct sunlight that location receives—and therefore the warmer the temperatures likely experienced at that time of year. The opposite also holds true: the smaller the elevation angle (the closer the sun is to the horizon at a given location), the less direct sunlight that location receives—and generally the colder that location is at that time of year.

Obliquity or tilt variations do not fundamentally change the annual amount of solar radiation received at a location, only the seasonal distribution of that sunlight. Changes over the course of Earth's full 41,000-year obliquity cycle vary from a minimum tilt of

21.5° to that of a maximum tilt of 24.5° and back. Currently, the axial tilt of Earth is close to 23.5° and is decreasing so that it will reach its minimum value around the year 10,000 CE.

The third of the orbital mechanisms is called the precession of the longitude of the perihelion in relation to the moving vernal (for the Northern Hemisphere) equinox, or more commonly, precession. In brief, this mechanism refers to the wobble of Earth's axis over time relative to the sun and is caused by the gravitational attraction on Earth by the other planets in the solar system, mainly Jupiter and Saturn. Two primary long-term cycles are associated with precession, that of 19,000 and 23,000 years. The precessional wobble changes the direction of Earth's axis of rotation relative to the fixed stars. Precessional wobble therefore changes the time of year in which perihelion, or the point in the orbit when Earth is closest to the sun, occurs, and also the time of aphelion, or the point in the orbit when Earth is most distant from the sun.

These three mechanisms form the core of a theory that dates back to the mid-1800s. In 1842, a theory was proposed that changes in these long-term orbital features could be the primary factors in establishing an ice age on Earth.[2] This idea linked to geologic work that indicated our planet has gone through a series of long-term cold events—the so-called ice ages—over the last two million years in which large continental ice sheets developed and grew. The key factor cited by the early theorists into how orbital changes could lead to ice ages was the seasonal distribution of solar radiation. The theorists suggested that a series of short, hot summers coupled with long, cold winters could produce the kind of conditions that are necessary to start an ice age. Today, we theorize the opposite; the critical control on ice age development likely involves a succession of long, cool summers and short, mild winters.[3]

The problem for these early theorists in the 1800s is that they didn't have the mathematical precision in their astronomical calculations of Earth's orbit and character to provide firm support to the theory. So the debate continued into the twentieth century. Then in 1941, the Serbian astronomer Multin Milankovitch published a set of detailed calculations of Earth's orbital mechanisms extending back a million years and the corresponding solar radiation received on Earth because of those mechanisms.[4] He computed these values for the high latitudes of the Northern Hemisphere—a region long thought to be critical to the formation, development, and eventual contraction of the great continental ice sheets needed for an ice age. Milankovitch discovered that a simple relationship appeared to exist between periods of low summer solar radiation and the existence of the four extensive European ice ages, known at that time to exist during the previous six hundred thousand years (now we know there have been many more).[5]

Following Milankovitch's work in the 1940s, research during the next three decades concentrated on establishing geologic evidence to support his hypothesis that orbital changes forced our Earth into and out of ice ages. In particular, the science involving marine deep-sea ocean core analysis was born and began to flourish. The vast majority of those deep-sea ocean cores indicated variations in climate that matched the solar radiation changes computed by Milankovitch.

By the late 1970s, the development of computer technology allowed for much more precise and detailed solar radiation calculations based on those orbital changes. In particular, a very brilliant Belgian scientist named André Berger wrote an involved computer program that carefully accounted for Earth's orbital changes with a much higher precision than was achieved by Milankovitch.[6] These radiation calculations, together with the extensive geologic proxy records from the deep-sea marine cores, were linked together by the powerful cycle-identifying statistical test termed *spectral analysis* (as discussed in the Mystery of the Great American Dust Bowl).

When scientists applied spectral analysis to the available deep-sea core data in the late 1970s, they discovered four strong underlying climate cycles, specifically cycles at around 100,000 years, 41,000 years, 23,000 years, and 19,000 years. According to these scientists, those specific cycles in paleoclimate data could not exist without being fundamentally linked to the basic pulses of the orbital mechanisms (eccentricity, 100,000 years; obliquity, 41,000 years; and precession, 23,000 and 19,000 years).[7] The correspondence was beyond coincidence. They concluded that the past climate variations of the Earth had to be directly tied to orbital changes of our planet. In fact, the three scientists, J. D. Hays, John Imbrie, and Nicholas Shackleton, published those results in a landmark paper in the world's premier technical journal, *Science*, and demonstrated the statistical link between our Earth's orbit and climate change. Despite the normal multisyllabic pedantry of scientists, the title Hays and his colleagues chose for their article was actually rather poetic: "Variations in the Earth's Orbit: Pacemaker of the Ice Ages."[8]

The problem was that fundamentally the only thing they had identified was a statistical linkage between orbital changes and climate—that the underlying cycles between the two data sets appeared to coincide. The physical explanation between the orbital changes was a bit more difficult to create. Because the computed changes in the solar input caused by the orbital mechanisms were very small, they didn't account for the amount of energy needed to produce and then melt huge continental ice sheets. While the major cycles between the orbital mechanisms and past climate seemed to coincide, there had to be something more to link changes in the orbital mechanisms to past climate change.

That something more was feedback. As discussed in the first chapter, feedback is based on the concept that when something is done to the climate system, the system itself can cause that something to be greatly amplified or depressed. A simple example of feedback involves an electric guitar and amplifier. When you strike a chord near the amplifier, the resulting sound out of the amplifier is a piercing shriek—because the sounds of both the guitar and the amplified sound continue to loop through the amplifier until they build to that ear-splitting screech. The basic nature of feedback is to amplify a small change into something bigger (although feedbacks can also act to dampen a change to the system).

A classic case of feedback in climatology is the *snow-temperature feedback*. It works like this: Suppose something causes temperatures globally to become slightly colder. If that happens, slightly more snow falls across our Earth. Snow has a higher reflectivity, or *albedo*, than does bare ground, so more sunlight is reflected back into space. If more sunlight escapes back into space, then Earth becomes colder—and the cycle will repeat, continually amplifying the cold.

Of course, the feedback can work the other way as well. If Earth becomes slightly warmer, then less snow falls globally and more sunlight is absorbed by the bare ground. More sunlight on the ground means warmer temperatures—and that cycle of increasing warmth will amplify over time.

The trouble is that there are hundreds, perhaps thousands, of feedbacks operating in our climatic system. For instance, the example above is a feedback that can amplify global warming; another one, termed the *evaporation-cloud feedback*, can diminish it. Suppose something causes Earth to warm up. If that is the case, then more ocean water evaporates under the warmer temperatures. But the loading of more water into the air leads to more clouds, which reflect sunlight back into space. Less sunlight at the surface produces a colder surface—thereby a continual diminution of the original warming signal.

Much of the work of the last half of the twentieth century in climatology was involved with identifying and evaluating different climate feedback processes in our atmosphere. If the snow-temperature and the evaporation-cloud feedbacks are both operating, for example, which one dominates? However, the end result of all that research remained the same; small changes in the amount of sunlight received on Earth could produce large changes in climate due to amplifying feedbacks. Milankovitch was right. His carefully calculated orbital theory could explain the occurrence of the ice ages of the last two million years.

The hundreds of deep-sea marine sediment core scientists loved this orbital cycle theory. Those scientists had a conceptually sound, working theory to which they could tie their carefully procured (and expensive!) deep-sea samples. Indeed, in the 1980s, most scientists involved with climate change were delighted with Milankovitch's astronomical theory and a group of them published a massive two-volume compendium documenting just how extensive the evidence supporting Milankovitch's orbital forcing of climate actually was.[9] But, in the 1990s—and remember science is not a democracy; the majority opinion doesn't win, it just gets the lion's share of federal funding—a scientist by the name of Isaac Winograd, together with a number of colleagues, had a really big problem with Milankovitch's orbital theory.[10] They said rather definitively that it simply didn't work.

The key to long-term scientific approval is whether Winograd and his team could back up that assertion with good data. Unfortunately, for the deep-sea marine scientists, Winograd's data were some of the very best.[11] They were working at a very isolated, tiny sinkhole located a hundred miles or so from Las Vegas, Nevada (yes, they were also working on understanding the long-term climate of

Yucca Mountain). That sinkhole is called Devil's Hole and, before its climatic fame, it was known for the minute endangered pupfish that were only found in its inaccessible waters.

The climate record of Devil's Hole rocked the climate world. Winograd had unearthed a record of past climate variations in its calcite walls that was better in quality than any of the hundreds of deep-sea marine cores taken around the world. And when he compared the climate changes seen in his Devil's Hole samples to orbital theory's calculated solar radiation changes, he found that the two records didn't *quite* match up. Consequently, Winograd and his colleagues asserted on the strength of their isolated, single location that all of the deep-sea marine cores around the world were wrong—that Milankovitch's orbital variations did not explain past climates.

Those were fighting words to the deep-sea marine core scientists. Winograd was implying that everything they had done with their expensive deep-sea cores was wrong. But the problem was that they couldn't show that his data were wrong. How could this situation be addressed?

Dr. John Shaffer (left) and the author standing at the Devil's Hole sinkhole, famous for its long, reconstructed climate record, in southern Nevada.

It turned out that the problem was simply comparing apples and oranges. The deep-sea marine sediment cores were measuring the variations in the growth and decline of the huge continental ice sheets that covered northeast North America and northern Europe—and, consequently, the solar radiation at those high latitudes (specifically at 60° North latitude). Conversely, Winograd's Devil's Hole climate record was responding to the rainfall and evaporation that occurred in the Desert Southwest (specifically at 30° North latitude). Perhaps the sunlight received at those different records was markedly different?

A talented former graduate student of mine, Dr. John Shaffer, showed the world in an article in the respected journal *Geology* exactly that—and carried it further.[12] He found that if a person computed solar radiation at specific latitudes and for different times of the year using the Milankovitch calculations, those irksome differences between Winograd's Devil's Hole data and the deep-sea marine core records could be reconciled. The Devil's Hole climate record was responding to slightly different solar radiation totals than those of the deep-sea marine sediment cores. Milankovitch's orbital theory was saved. The available evidence supported the conclusion that past climates—at least for the last two million years or so—have been primarily determined by changes in the orbital mechanisms of Earth.

Unfortunately for climate reconstructionists, by the turn of the new millennium, nobody outside of the paleoclimate world cared whether Milankovitch's climate change theory was valid or not. Global warming had entered the scene, and warm was trumping cold.

A key point here, however, is that astronomical theory is still credible—and is applicable to the future! We can actually calculate future orbital changes of Earth and thereby determine the likely climate change generated by those orbital variations. What we discover when we make those calculations is that our planet's orbital changes are slowly forcing Earth to colder conditions. If not a full-blown ice age, the changes in Earth's orbit will definitely produce a very intense Little Ice Age about ten thousand years from now.

But wait a minute! What about global warming? We have politicians, the media, and many scientists broadcasting alarms that we are headed down an irreversible path of rising seawaters and hotter temperatures. How can we possibly have a near ice age (geologically speaking) in a few short thousand years? Global warming is occurring, isn't it?

I always am perplexed by members of the media (and politicians) who harp on that question, as if to answer it no means you are an oil company lackey and should be completely disregarded. But to answer yes—which is the right answer, with qualifications—must validate the great public concerns with global warming. A true black-and-white, either-or situation. Sorry, but as with so many things in this world, I don't think the global warming climate problem is quite so clear-cut. I believe that to understand the ramifications of global warming, one must have something that has generally been lacking in the heated public debate on global warming—a historical perspective.

And the historical perspective is that global warming will be (relatively speaking again) short term. By that I mean, the most liberal estimates for untapped fuel reserves is perhaps three hundred years. After that period, if we can't put hydrocarbons into the air, then global warming will gradually diminish—unless we come up with an equally climate-altering alternative.

So the processes that influence us in the very short term (I'm speaking of decades), such as global warming through human-made pollutants, are not likely to be the major processes influencing us in the long term (such as centuries and longer). For Earth over at least the last couple of million years, the critical mechanisms that influenced long-term climate change have been the orbital changes of our planet. And those orbital changes suggest that Earth is likely to get colder in the long term—enough to force the planet into at least another severe Little Ice Age.

There remains, however, the possibility of a couple of jokers in the deck that prevent me from offering that ten-thousand-year forecast unconditionally. I cautioned in the Mystery of the Devil in the Deep Blue Sea that the oceans—and their devilish "conveyor belt" of salt and heat—are critical to long-term climate change. Unfortunately, we have likely much to learn about how the global thermohaline circulation of the oceans actually works—and why. So any long-term forecast must be hedged with extensive caveats about our knowledge of the oceans' influence on climate.

Second, we have to always be on guard for the possibility of catastrophes. Mysteries such as that of the Dead T. Rex (resulting from an asteroid colliding into Earth), the Little Ice Age (resulting from the cessation of sunspot activity on the sun), and Humanity's "Near Extinction" (resulting from the eruption of supervolcano Toba) point out that Earth's climate can shift abruptly due to a major change that can't be predicted. Any long-term forecast must be able to give some credence to the potential (even if small) occurrence of such events.

But, wait a minute. I didn't mention any climate model for my ten-thousand-year forecast. Aren't climate models supposed to be our best means to study future climate change?

A climate model is a fundamental tool for the climatologist. In that regard, it is similar to any other type of weather equipment such as a rain gauge or a thermometer. A climate model is meant to help us understand climate. But, just like any tool, it cannot be used for every purpose. Since I built a general circulation climate model as part of my doctoral dissertation research many years ago, let me discuss climate models in a bit more detail and show some of the strengths—and weaknesses—of climate models.

Changes in weather and climate can be expressed in specific mathematical terms as the result of the forces and the effects influencing them. We know, for example, that the changes in the winds that we experience across the planet are the result of four specific factors: (a) first and foremost, the differences in air pressure between disperse locations, (b) the changes in direction of flow resulting from Earth's rotation (the Coriolis effect), (c) friction, the force that retards motion, and (d) gravity. We can express those factors as mathematical expressions: if we know the values of those expressions, we can compute changes in wind speed and direction over time—and make forecasts.

All of the critical properties of our atmosphere can be expressed in (surprisingly!) only five basic mathematical equations, called the *primitive equations* (not because of any simplicity but because of their primal characterization of our atmosphere), or *fundamental equations*.[13] The primitive equations include: an equation that computes wind speeds and directions using the forces discussed above (*Navier-Stokes equation*); an equation that equates atmospheric pressure, density, and temperature (*equation of state*); a math equation that accounts for all of the air in our atmosphere (*equation of mass conservation*); an equation addressing the causes of temperature variations (*thermodynamics relationship*); and a fifth mathematical equation addressing the variations in water vapor (*equation of moisture conservation*). These five equations contain five "unknowns," variables whose future values we are trying to determine: pressure, density, temperature, wind, and moisture. In mathematics, if we have the exact number of equations as we do the number of unknowns, then we should explicitly and precisely be able to solve those equations—a process that mathematicians term mathematical closure. For example, if I have two variables, x and y, and two equations: $x + y = 3$ and $x = 1$, then mathematical closure dictates that I can solve them. Indeed, in this simple example, the solution would be $y = 2$.

Consequently, from a mathematical viewpoint, with the closure of the five primitive equations, we should have perfect forecasts.

Unfortunately, it isn't quite that simple. There are two big concerns. First, those five primitive equations are termed *second-order nonlinear partial differential equations* and they require some high-powered calculus to solve. More important, they can't be solved explicitly on a computer. Perhaps surprisingly to some, computers don't do calculus. They add, subtract, multiply, and divide very well (and very quickly), but they don't solve calculus equations of this type. So we have to rewrite the equations into a form that computers can handle—and that contributes to error. We need to simplify, or perhaps "approximate" is a better word, the primitive equations by developing shortcuts—roundabout ways of estimating some of the mathematical terms, creating what are called *parameterizations*. But, because of that, a climate model is therefore only as good (or as bad) as its parameterizations. One of the big changes throughout the last thirty years of computer climate model development has been in the increasing complexity (and reality) of the parameterizations used in the models.

Second, the solution to the primitive equations requires a lot of detailed input information (the starting values of those five unknowns) across Earth—and at a multitude of levels in the atmosphere. Unfortunately, large parts of our planet are uninhabited and relatively unknown. Determining the starting model values for those locations involves some rather uncertain estimations and large assumptions about how representative of their surroundings those starting values actually are. (Recall the Mystery of the Crashing Airplane and the A-Bomb, where I discussed the GIGO law: Garbage In, Garbage Out.) The quality of the results from computer climate models is always and inexorably constrained by the quality of the initial data fed into them.

So, taking into account those two big concerns, how good are climate models?

Actually climate models are very good if used for what they were designed to do. Fundamentally, computer models are tools to gauge how complete our understanding of the atmosphere actually is. By running them and comparing them to their observed climate counterparts, and to the results from other climate models, we begin to see how accurate the model parameterizations of specific parts of our atmosphere are. We can, therefore, start to assess how sensitive certain parts of our climate system are to overall changes, according to the physics principles that we currently understand.

A climate model simulation tells us how well we have replicated the world. Recall my early discussion of climate mysteries in which the initial research on nuclear winter by the TAPPS group was subsequently reevaluated in relation to climate feedbacks.[14] We learned through climate feedbacks that doing something to our environment could cause other somethings that somehow led to still other somethings that would either eventually amplify or diminish the original changes. The follow-up scientists to the TAPPS study discovered that the original computer model for the nuclear war simulations lacked some of these critical feedback processes.[15]

So a climate model allows us to study the quality of the parameterized components of our conceptual atmosphere. Just like a model plane can be used in a wind tunnel to test its aerodynamics, a climate model can be used to test how our climate system works—but with the understanding that its workings depend on how accurately the model parameterizations represent reality. In the case of a model plane, the analogue might be how well the model plane has been built using glue and wood to represent the rivets and metal of the real plane.

Climate models were not—and for the most part, are still not—designed to make forecasts, just like model planes are not designed to haul cargo. Most climatologists are aware that climate models provide us with critical climate information that we often can't get any other way (such as our more detailed understanding of the Age of the

Dinosaurs as discussed in the Mystery of the Dead T. Rex), but they aren't designed specifically to forecast conditions—to perfectly replicate all aspects of the climate system—for long times into the future.

It may seem like a trivial distinction, but it is important. Climate models are only as good as our current understanding of the climatic system. Thirty years ago, climate models consisted of mathematical worlds that contained "oceans" that were a half an inch deep, that were "dead" in terms of interactive vegetation, and that didn't even have day/night cycles. Our understanding of climate has improved over the last thirty years, as have the speed and accuracy of computers, so we are creating better and better climate models—meaning, they are becoming closer and closer representations of the actual climate system.

Consequently, the question that must be asked—and this is critical in today's climate debates—is simply "At what point can we accept a climate model's results as representative of reality?"

I wish I could give a definitive answer. To some extent, it becomes a personal choice. Does a person buy an insurance policy based on an actuary analysis of a random dozen people or does she buy an insurance policy based on a detailed statistical analysis of a random million people? If you say that you prefer the million-person model, then you're saying that a more accurate representation of the real world is critical—but at what exact point do we reach that representation? At a thousand? At a hundred thousand?

The same idea holds for climate models. The amount of credence that we can give to model findings is based on how accurately they parameterize the myriad of key elements of our climate system. Even though I said that there are only five basic mathematical equations defining our atmospheric system, the number of factors that influence the five basic variables in those five equations is enormous. Climate models have greatly progressed—become more accurate—over time; they have easily become some of the most complex computer programs ever written and, correspondingly, require the biggest supercomputers to run them. They now include detailed ocean models, the response of living plants and organisms, and even the growth and decay of sea ice. But they are not perfect: they are not exact reproductions of our climate.

At this time, the degree of faith that we place in computer models must therefore be determined via indirect means. Climatologists use comparisons as one for determining quality. For example, the closer the similarity among different climate models' results likely indicates a greater degree of certainty of the underlying physics.[16] In a similar fashion, the amount of correspondence between the models and reality (measured climate) is of critical importance.

Undoubtedly, our climate models will continue to improve. As they do so, our ability to eventually "predict" future climates of our planet will improve as well.

But at this time, I am tempted to remind everybody of how far we have come. A few thousand or even a few hundred years ago, if I had given a forecast for the next ten thousand years, there is a distinct likelihood of my ultimate fate as a seer being rather bleak. Those in history who have attempted to see into the future have sometimes suffered rather traumatic outcomes. Prophets like Nostradamus or Cassandra (as seen in this chapter's outset) were often not highly regarded by many of their compatriots. As we continue to learn more and more about our climatic system, it is my hope that our climate forecasts will continue to improve and thus be viewed with increasing favor.

Time: June 1566
Location: Paris, France, Europe
(This vignette is loosely based in part on an open letter from Nostradamus to the Privy Councillor in 1566.)[17]

Yes, **the old man realized,** ***the great Michel de Nostredame was dying.*** **One didn't need to be a seer to know that his gout had transformed into something much more debilitating and deadly. He lay under the silk sheets of the extravagant bed that the Queen Consort of France had provided for him and waited for death to finally arrive.**

It is fitting, **the bearded man thought as he tried in vain to ignore the continuing pains throughout his body:** ***I will likely soon foresee my own death—within just a few short weeks, perhaps even days.***

All he had tried to accomplish throughout his life was the creation of a simple almanac, a tool to help

people in organizing their lives. Why did people need to believe that what he said was either ordained by the Lord God or, conversely, by the very Devil himself? He had been so careful to give his advice in the form of playful word games and varied syntax even to the point of using an assortment of languages other than his native French. And for that, he had been praised by some . . . and condemned by others; while the great Nostradamus, as he styled himself, had been exulted by royalty, he had been reviled by some in the Church.

Well, it was time that people fully understood who Michel de Nostredame was. With a fluid-filled, raw cough, he gestured for his secretary to approach.

The man knelt down next to the bed.

"Continue the letter to the Privy Councillor," the dying man rasped, "with this: 'I do but make bold to predict (not that I guarantee the slightest thing at all), thanks to my researches and the consideration of what judicial Astrology promises me and sometimes gives me to know . . .'"

His speech was cut off by another bout of excruciating coughing.

Finally the old man continued after a moment, "'Not that I am foolish enough to pretend to be a prophet . . .'"

Chapter 22
THE SEVEN SOLUTIONS

When I teach a university class on climate change, I try at the end of the semester to ensure that the basic concepts of the class are emphasized, even if the minute details are quickly forgotten. I therefore often phrase the questions on my final exams in a manner that demonstrates "things I want you to know even two or three years after this class is over!" So after reading these little mysteries about climate change throughout history, are there any fundamental messages—any basic solutions arising from these mysteries—that I would like you to remember long after this book is put back on the shelf? Let's take a quick look at seven overriding ideas found in our greatest mysteries of climate and weather history.

First, *climate does change*. As we have seen from the Age of the Dinosaurs, through early civilizations and the Middle Ages, and into the Little Ice Age and modern times, climate has changed, is changing, and, frankly, will continue to change. We've seen that it can happen in an instant (as in the Mystery of the Dead T. Rex) or it can take centuries (the Mystery of the Vanishing Harappans), but without a doubt, climate does change. To think that we live (or can live) in a "stable" climate is a complete falsity—at least throughout the history of our planet, climate change has been the rule much more than climate stability over the long term, as Richard Alley so aptly pointed out in the Mystery of the Devil in the Deep Blue Sea.

That finding is neither good nor bad. Change in and of itself doesn't have a value judgment. So climate change is not inherently good or bad. The concept of change can only be evaluated in how it impacts the specific individual or a society at large. Often I am interviewed by the media here in the arid Southwest about whether a given winter forecast of, for example, wetter-than-normal conditions will be either good or bad for us. The problem with making such a value judgment is that it can be both. A wet winter might be great for ski companies in the mountains of Arizona but it could spell ruin for tourism officials promoting the warm, clear skies of the desert. Change doesn't imply good or bad.

Second, *climate and weather investigations require unique research methods and even more unique scientists to employ those methods.* We have seen that to completely understand the myriad of climates that our Earth has experienced throughout its past, we must extract climate information from literally scores of unusual sources. To solve our historical mysteries of climate and weather, scientists have needed to analyze mud cores from the bottom of the oceans, ice samples from the remote regions of Antarctica, rock art from the arid interior of the Sahara, the telling rings from ancient trees in Europe and Mexico, and even the historical documents of Christopher Columbus and Charles Darwin.

And the scientists analyzing those data have been equally diverse in their backgrounds and preferred styles of research. The detectives of our mysteries have been researchers who hail from a wide-ranging group of climate- and weather-related sciences, including dendrochronology, geomorphology, palynology, glaciology, astronomy, geography, epidemiology, and even biblical history. The investigation styles themselves are diverse and varied, ranging from the classic deductive approach employed by researchers like Russ Vose (in the Mystery of "Eliminating the Impossible"), to the inductive style preferred by Ted Fujita (in the Mystery of the Crashing Airplane and the A-Bomb), to the observational storytelling of Ray

Henkel (in the Mystery of Climate, Cocaine, and the Storyteller). I think one can easily conclude that climate and weather sciences are currently two of the most diverse fields of study in the world.

That investigative diversity relates to our third solution: *Beware of the black-and-white syndrome.* For instance, in the Mystery of the Vanishing Harappans, I discussed the extreme concept called *environmental determinism*—where the physical environment exerts an absolute control on society and culture—and its modern equivalent, *anthropogenic determinism*—where some have suggested that society and culture are exerting absolute control on the environment. Both extreme positions fail to account for the incredible intricacy of Earth's environmental system.

One point that I like to stress is that our climate system is one of the most complex systems of which we know. As I noted in the Mystery of the Future, when a new generation of computers is created, some of the first programs that analysts use to test its limits are computer climate models. Such models—having to address variations in glaciers, oceans, land cover, air currents, and other multitude of processes that compose our environment—involve some of the most complex computer codes ever written. And we continually learn more about that intricate system of processes all of the time.

With an atmospheric system as complex as ours, it has been my experience that a single, clear-cut answer about climate or weather is seldom, if ever, correct. As we saw in the Mystery of the Disappearing Sun, the story can change as new facts are uncovered. There are, unfortunately, an infinite series of grays in the world of climate and weather research, where we would like to have black-and-white, easy-to-understand results. To me that inherent uncertainty is a challenge—I love the fact that we are constantly discovering new facts about our atmospheric system—but to policy makers (and some scientists as well), that continual uncertainty can be very frustrating.

Fourth, without question, *civilizations can be influenced by climate and weather*. The big problem is determining the degree to which a given civilization is influenced by the atmosphere. In each mystery, we've seen a civilization that has been impacted—sometimes more, sometimes less—by its climate or weather. Some civilizations struggled to overcome the limitations of their natural climate (for example, the inhabitants of Petra with their amazingly engineered water systems), while others failed to adjust adequately to changes—particularly when coupled with other problems such as disease (as the traumas faced by the Maya in the Mystery of the Mayan Megadrought). Sometimes the climate has appeared to enhance the stability of a society (for instance, in the Qing dynasty period of China), while in others, the fluctuations in climate have impacted them negatively, such as in Europe during the Little Ice Age or in Homer's ancient Greece.

Climate's precise influence on a society is something, I believe, that must be addressed on a case-by-case basis. In my view of history and climate, I haven't discerned a clear, consistent, one-size-fits-all pattern of climate's influence on society. As much as I enjoy reading some of the recent historical syntheses about climate and society (some even winning major prizes), I find the case exceptions to climate-society interactions to be more prevalent (and perhaps more important) than those used to show similarities. But, unfortunately, pointing out important individual differences in civilizations and their response to climate change doesn't tend to attract as much attention as proclaiming, for instance, that pronounced drought always leads to the downfall of a civilization.

Fifth, *global climate change isn't global!* That message is interesting in that it might seem paradoxical. But with very few exceptions, our planet does not behave as a single unit. Different regions of the world respond differently to changes. As we saw in the Mysteries of the Little Ice Age and the Complacent Empire, the largest climate change that we have ever seen in the modern history of humankind, the time called the Little Ice Age, ushered in a very cold climate to eastern North America and Europe. But that time coincided with a relatively wet and mild climate over China. As Charles Darwin noted, even a "very general cause" of major weather changes around the world, such as El Niño–Southern Oscillation, doesn't impact the entire world in exactly the same way. While the major 1997–98 El Niño event led to massive rains in California and the Southwest, it created a drought and forest fires in Australia. When you hear of a general "global" finding about climate change, remember that our weather and climate mysteries have shown us that our Earth simply doesn't respond as a single unit.

Sixth, *a key to surviving climate change is how quickly, effectively, and intelligently we adapt to potentially hazardous changes—and we need to keep an eye out for the associated problems.* At my university and others around the world, the current buzzword involving climate change is *sustainability*. Recall our solution to the Mystery of Petra, the Rose City. The inhabitants of that great desert city overcame their environmental constraints through a combination of innovation, money, and engineering.

With regard to current climate change, we've designed massive engineering projects on the theoretical drawing board that would redistribute heat, water, energy, and food across the entire globe if needed. These include climate engineering, or *geoengineering*, plans as diverse as damming the Gulf Stream or the Bering Strait, towing icebergs from Alaska to California, or even artificially generating oceanic algae blooms for carbon dioxide intake.[1] Of course, before we rush out and solve the world's climate problems, there exists a fundamental problem with such massive climate engineering ventures. Simply stated, we absolutely must understand the consequences intended and unintended of such projects.[2] We have only one Earth right now to mess with. Once we start to tinker with an incredibly complex system such as our climate system, there is no turning back.

An important corollary to this point, at least in my understanding of history and climate, is that a large part of civilizations' problems with climate change haven't been with the climate change itself, but with a multitude of related tribulations. The Mystery of the Mayan Megadrought, for example, suggested that the critical element of the Mayan collapse was in all probability a killer disease, not the drought itself. As Dave Stahle and his fellow researchers noted, a megadrought may have contributed to the epidemic outbreak by giving it a proper environmental setting, but the megadrought didn't fundamentally cause the outbreak. In any discussion of future climates, we must identify all of the potential factors associated with climate change and be prepared to address those factors—as well as the climate change itself—quickly, intelligently, and forcibly.

And finally, my seventh and last fundamental message is that *climate will likely stay mysterious far, far into the future*. I noted at the outset of this book the peculiar newness of climate and weather research. I am a scientist in a field where new discoveries happen almost daily. It is amazing to realize after more than five thousand years of recorded human history, we are still discovering new kinds of weather and climate (for instance, the microbursts of the Mystery of the Crashing Airplane and the A-Bomb). That vast amount of unknowns in climate science and meteorology can be daunting in demonstrating to us just how much of our weather and climate likely remains a mystery to us. We have so much to learn about our climate system that I can only imagine what a climate textbook in only ten or twenty years from now will encompass. With discoveries only within the last few decades of new forms of lightning (*elves*, *sprites*, and *blue jets*), of new climate teleconnections (such as the Pacific Decadal Oscillation), of weather in new and unusual places (for instance, the incredibly rare hurricane that hit South America in 2001), we continue to see how much we have yet to learn. But, while that lack of knowledge is perhaps frustrating to many as we struggle with important climate and weather policy issues, it is also exhilarating for scientists, and perhaps intriguing for others, in that it shows us how wonderful, complex—and mysterious—our atmosphere truly is.

Time: Fall 2010
Location: A classroom at Arizona State University, Tempe, Arizona, North America

"All right, we'll continue with that thought in our next class. Oh, be sure to read pages 102 to 142 that are listed on the syllabus by next Tuesday!" With a smile, the redheaded climatology professor standing at the central podium of the lecture hall waved a general dismissal to the large mass of students.

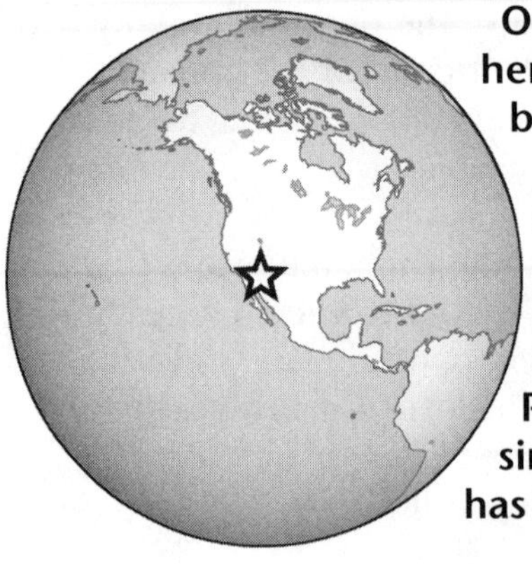

One of the new freshmen turned to her friend as they gathered up their books and noted.

"This climate and weather stuff is weird, if you ask me! I mean, climate detectives? Who'd have thought that they've dug into the ice down at the South Pole or chopped bits out of trees simply to tell us what the weather has been like thousands of years ago."

Her friend nodded. "But, you know, I've been thinking that it's kind of exciting—a lot more than in some of my other classes, anyway. And I hadn't realized that a guy could make a career out of studying weather and climate other than being the local TV weatherman or working for the weather service."

The student turned to her friend in surprise. "Are you actually thinking of changing your major?"

"Well, these climate and weather mysteries do sound kind of interesting . . . and important. I don't know if I'll change majors right now, but it's something to think about anyway."

The two students exited the lecture hall deep in conversation about the merits of a possible climate or weather career.

The redheaded climatology professor smiled with satisfaction as he watched them depart.

ACKNOWLEDGMENTS

As you have read through this book, you have probably become aware that I owe enormous debts of gratitude to a very large number of people. First and foremost, of course, is my family and particularly my dear wife, Niccole, who put up with me over the course of this book's writing. Second, this book simply couldn't have been written without the amazing research of some absolutely brilliant people whom I have had the distinct pleasure of interacting with throughout my twenty-plus years in climate research. This includes exceptional people like the individual who first perked my interest in this subject so many years ago and who went on to be the best mentor a young scientist could have, Dr. Merlin Lawson; brilliant colleagues such as Dr. Bob Balling, Dr. Ron Dorn, Dr. Tony Brazel, Dr. Jay Hobgood, Dr. Richard Alley, Dr. Ray Henkel, Dr. Pat Fall, and the late Dr. Mel Marcus; accomplished fellow scientists such as Rex Adams and Ron Holle; and quite simply some of the most talented students that a professor could wish for—former students like Dr. David Stahle, Dr. John Shaffer, Dr. Russ Vose, Dr. Nancy Selover, Dr. David Brommer, and Dr. Grady Dixon as well as current students such as Bo Svoma, Matt Pace, Kimme Debiasse, and Jessica Nolte. I also appreciate the brilliant people with whom I have collaborated in many of my public outreach articles including such people as Dr. Joe Schaefer, Drs. Charles and Nancy Knight, Dr. Chris Landsea, Dr. Roger Edwards, Dr. John Harrington, Dr. Tom Peterson, and Dr. Pierre Bessemoulin.

But beyond those gifted researchers, I have been blessed with the absolute best technical support for this book; people like Becky Eden, who did the wonderful cartographic globes used in each chapter, and the extraordinarily talented Barbara Trapido-Lurie, a gifted cartographer and good friend who created many of my graphics throughout the book (and indeed throughout my entire research career!). I owe a huge debt of gratitude to the many old and new friends who kindly contributed the many beautiful photographs used throughout the book; people like Dr. Tom Paradise of the University of Arkansas, Dr. Andrea Zerbon of the University of Milan, Dr. Mark Hildebrandt of the University of Illinois–Evanston, Dr. Caroline (Molly) Davies of the University of Missouri–Kansas City, Drs. Pat Fall, Mike Kuby, Elizabeth Larson-Kaegy, and Tony Brazel of Arizona State University, Ron Holle, lightning expert extraordinaire, the "illuminating" solar photographer Greg Piepol of www.sungazer.net and, last but not least, the incomparable Dr. Ray Henkel. And, of course, I am indebted to my fabulous agent, Andrée Abecassis—having as talented of an agent as Andrée makes life so much easier—and to the professionals at Prometheus Books, particularly Linda Regan.

Beyond those who helped with the research, writing, and creation of the book lies a huge number of friends in the media who have given me the opportunity of putting some complex weather science into understandable terms; people such as the incredible folks at the Heldref magazine *Weatherwise* from the late great editors David Ludlum and Pat Hughes, to its current gifted captain, Margaret Brenner, Arizona's own media legend Pat McMahon (and his extremely talented producer Laura Holka), Ed Phillips, Royal Norman, Sean McLaughlin, Steve Garry, John Hook, Dave Munsey, Bill Bellus, Jeff Rosenfeld, Keith Jennings, Skip Derra, Jim Cross, Doyle Rice, the great and gifted people at the Weather Channel,

Steve Koppes at the University of Chicago, and the multitude of people who have taken the time to drop me a line with something to add to my weird weather archive or chat about the oddities of climate and weather.

Finally, I must acknowledge my debt to a man whom I regretfully never had the chance to actually meet in person, but who without question has influenced my life greatly. Robert A. Heinlein's thoughts and ideas have sparked my interest in the world since I secretly read my first Heinlein book during a seventh-grade English class (which consequently might have contributed to some of the grammatical mistakes I have probably made in this book). I might say more, but as the dean of science fiction himself wisely advised, "Don't try to have the last word. You might get it."[1]

Randy Cerveny
President's Professor of Geographical Sciences
Arizona State University
February 2009

CHAPTER 1: WHAT IS A WEATHER MYSTERY?

1. R. A. Heinlein, *Time Enough for Love* (New York: G. P. Putnam's Sons, 1973), p. 352.

2. R. P. Turco, O. B. Toon, T. P. Ackerman, J. B. Pollack, and C. Sagan, "Nuclear Winters—Global Consequence of Multiple Nuclear Explosions," *Science* 222 (1983): 1283–92.

3. S. H. Schneider and S. L. Thompson, "Simulating the Climatic Effect of Nuclear-War," *Nature* 333 (1988): 221–27.

4. G. Santayana, *Reason in Common Sense*, vol. 1 of *The Life of Reason: or, The Phases of Human Progress* (New York: C. Scribner's Sons, 1906), p. 285.

5. B. D. Shaw, "Climate, Environment, and History: The Case of Roman North Africa," in *Climate and History: Studies in Past Climates and Their Impact on Man*, ed. T. M. L. Wigley, M. J. Ingram, and G. Farmer (Cambridge: Cambridge University Press, 1981), p. 380.

6. E. Huntington, *The Character of Races as Influenced by Physical Environment, Natural Selection and Historical Development* (New York: C. Scribner's Sons, 1924).

7. R. S. Cerveny and R. C. Balling, "Identification of Anthropogenic Weekly Cycles in Northwest Atlantic Pollution, Precipitation and Tropical Cyclones," *Nature* 394 (1997): 561–62.

8. S. A. Changnon Jr., F. A. Huff, and R. G. Semonin, "METROMEX: An Investigation of Inadvertent Weather Modification," *Bulletin of the American Meteorological Society* 52 (1971): 958–67.

9. T. L. Bell, D. Rosenfeld, K. M. Kim, J.-M. Yoo, M. I. Lee, and M. Hahnenberger, "Midweek Increase in US Summer Rain and Storm Heights Suggests Air Pollution Invigorates Rainstorms," *Journal of Geophysical Research—Atmospheres* 113(D2) (2008): D02209.

10. P. M. D. Forster and S. Solomon, "Observations of a 'Weekend Effect' in Diurnal Temperature Range," *Proceedings of the National Academy of Sciences* 100 (2003): 11225–30.

CHAPTER 2: THE MYSTERY OF THE DEAD T. REX

1. E. J. Barron and W. M. Washington, "The Role of Geographic Variables in Explaining Paleoclimates: Results from Cretaceous Climate Model Sensitivity Studies," *Journal of Geophysical Research—Atmospheres* 89 (1984): 1267–79; L. C. Sloan and E. J. Barron, "Equable Climates during Earth's History," *Geology* 18 (1990): 489–92.

2. A. M. Agustsdottir, E. J. Barron, K. L. Bice, L. A. Colarusso, J. L. Cookman, B. A. Cosgrove, J. L. De Lurio, J. F. Frakes, L. A. Frakes, C. J. Moy, T. D. Olszewski, R. D. Pancost, C. J. Poulsen, C. M. Ruffner, D. G. Sheldon, and T. S. White, "Storm Activity in Ancient Climates 2. An Analysis Using Climate Simulations and Sedimentary Structures," *Journal of Geophysical Research—Atmospheres* 104 (1999): 27295–320.

3. Ibid.

4. A. R. Hildebrand, G. T. Penfield, D. A. Kring, M. Pilkington, A. Camargo, S. B. Jacobsen, and W. V. Boynton, "Chicxulub Crater: A Possible Cretaceous/Tertiary Boundary Impact Crater on the Yucatan Peninsula, Mexico," *Geology* 19 (1991): 867–71.

5. L. W. Alvarez, W. Alvarez, F. Asaro, and H. V. Michel, "Extraterrestrial Cause for the Cretaceous-Tertiary Extinction: Experimental Results and Theoretical Interpretation," *Science* 208 (1980): 1095–1108; D. A. Kring, "Air Blast Produced by the Meteor Crater Impact and a Reconstruction of the Affected Environment," *Meteoritics and Planetary Science* 32 (1997): 517–30.

6. Alvarez, Alvarez, Asaro, and Michel, "Extraterrestrial Cause for the Cretaceous-Tertiary Extinction."

7. B. O. Bressler and W. U. Reimold, "Terrestrial Impact Melt Rock

and Glass," *Earth-Science Reviews* 56 (2001): 205–84; Kring, "Air Blast Produced by the Meteor Crater Impact and a Reconstruction of the Affected Environment."

8. M. A. Kruge, B. A. Stankiewicz, J. C. Crelling, A. Montanaria, and D. F. Bensley, "Possible Charcoal in Cretaceous-Tertiary Boundary Strata: Evidence for Catastrophic Firestorm and Megawave," *Geochimica et Cosmochimica Acta* 58 (1994): 1393–97.

9. P. J. Crutzen, "Mass Extinctions: Acid Rain at the K/T Boundary," *Nature* 330 (1987): 108–109.

10. J. Bouregeois, T. A. Hansen, P. L. Wiberg, and E. G. Kauffman, "A Tsunami Deposit at the Cretaceous-Tertiary Boundary in Texas," *Science* 241 (1988): 567–70; Kruge, Stankiewicz, Crelling, Montanaria, and Bensley, "Possible Charcoal in Cretaceous-Tertiary Boundary Strata"; H. J. Meloshi, N. M. Schneider, K. J. Zahnle, and D. Latham, "Ignition of Global Wildfires at the Cretaceous/Tertiary Boundary," *Nature* 343 (1990): 251–54.

11. Hildebrand, Penfield, Kring, Pilkington, Camargo, Jacobsen, and Boynton, "Chicxulub Crater."

12. N. Van Sandick, 1884 quote in J. R. Nash, *Darkest Hours* (Chicago: Nelson Hall, 1976).

13. P. J. Crutzen, "Mass Extinctions"; C. Covey, S. L. Thompson, P. R. Weissman, and M. C. MacCracken, "Global Climatic Effects of Atmospheric Dust from an Asteroid or Comet Impact on Earth," *Global and Planetary Change* 9 (1994): 263–73.

14. Covey, Thompson, Weissman, and MacCracken, "Global Climatic Effects of Atmospheric Dust from an Asteroid or Comet Impact on Earth."

15. B. Lomax, D. Beerling, G. Upchurch, and B. Ott-Bliesner, "Rapid (10-Yr) Recovery of Terrestrial Productivity in a Simulation Study of the Terminal Cretaceous Impact Event," *Earth and Planetary Science Letters* 192 (2001): 137–44.

CHAPTER 3: THE MYSTERY OF HUMANITY'S "NEAR EXTINCTION"

1. S. A. Tishkoff, E. Dietzsch, W. Speed, A. J. Pakstis, J. R. Kidd, K. Cheung, B. BonneTamir, A. S. Santachiara-Benerecetti, P. Moral, M. Krings, S. Paabo, E. Watson, N. Risch, T. Jenkins, and K. K. Kidd, "Global Patterns of Linkage Disequilibrium at the CD4 Locus and Modern Human Origins," *Science* 271 (1996): 1380–87.

2. S. H. Ambrose, "Late Pleistocene Human Population Bottlenecks, Volcanic Winter, and Differentiation of Modern Humans," *Journal of Human Evolution* 34 (1998): 623–51.

3. C. F. Miller and D. A. Wark, "Supervolcanoes and Their Explosive Supereruptions," *Elements* 4 (2008): 11–15.

4. M. R. Rampino and S. Self, "Volcanic Winter and Accelerated Glaciation following the Toba Super-Eruption," *Nature* 359 (1992): 50–52.

5. M. R. Rampino and S. Self, "Historic Eruptions of Tambora (1815), Krakatau (1883), and Agung (1963), Their Stratospheric Aerosols, and Climatic Impact," *Quaternary Research* 18 (1982): 127–43.

6. EPICA Community Members (56 authors), "Eight Glacial Cycles from an Antarctic Ice Core," *Nature* 429 (2004): 623–28.

7. Rampino and Self, "Volcanic Winter and Accelerated Glaciation following the Toba Super-Eruption."

8. C. Lang, M. Leuenberger, J. Schwander, and S. Johnsen, "16°C Rapid Temperature Variation in Central Greenland 70,000 Years Ago," *Science* 286 (1999): 934–37.

9. M. T. Jones, R. S. J. Sparks, and P. J. Valdes, "The Climatic Impact of Supervolcanic Ash Blankets," *Climate Dynamics* 29 (2007): 553–64; M. R. Rampino, "Supereruptions as a Threat to Civilizations on Earth-Like Planets," *Icarus* 156 (2002): 562–69.

10. R. B. Smith and L. W. Braille, "The Yellowstone Hotspot," *Journal of Volcanology and Geothermal Research* 61 (1994): 121–87.

11. Jones, Sparks, and Valdes, "The Climatic Impact of Supervolcanic Ash Blankets."

CHAPTER 4: THE MYSTERY OF THE SAHARAN HIPPOS

1. N. V. Cerveny, R. Kaldenberg, J. Reed, D. S. Whitley, J. Simon, and R. I. Dorn, "A New Strategy for Analyzing the Chronometry of Constructed Rock Features in Deserts," *Geoarchaeology* 21 (2006): 281–303.

2. D. Whiteley, ed., *Handbook of Rock Art Research* (Walnut Creek, CA: AltaMira Press, 2001).

3. F. Mori, R. Ponti, A. Messina, M. Flieger, V. Havlicek, and M. Sinibaldi, "Chemical Characterization and AMS Radiocarbon Dating of the Binder of a Prehistoric Rock Pictograph at Tadrart Acacus, Southern West Libya," *Journal of Cultural Heritage* 7 (2006): 344–49.

4. R. I. Dorn, D. B. Bamforth, T. A. Cahill, J. C. Dhrenwend, B. D. Turrin, D. J. Donahue, A. J. T. Jull, A. Long, M. E. Macko, E. B. Weil, D. S. Whitley, and T. H. Zabel, "Cation-Ratio and Accelerator Radiocarbon Dating of Rock Vanish on Mojave Artifacts and Landforms," *Science* 231 (1986): 830–33; R. I. Dorn, P. B. Clarkson, M. F. Nobbs, L. L. Loendorf, and D. S. Whitley, "New Approach to the Radiocarbon Dating of Rock

Varnish, with Examples from Drylands," *Annals of the Association of American Geographers* 82 (1992): 136–51.

5. W. Schott, *Deep-Sea Sediments of the Indian Ocean* (Tulsa: American Association of Petroleum Geology, 1939), pp. 396–408.

6. H. H. Lamb, *Climate, History and the Modern World*, 2nd ed. (London: Routledge, 1995).

7. K. W. Butzer, *Environment and Archeology* (Chicago: Aldine Publishing, 1964).

CHAPTER 5: THE MYSTERY OF THE VANISHING HARAPPANS

1. E. Huntington, *Civilization and Climate* (New Haven, CT: Yale University Press, 1924).

2. A. Brazel, P. Gober, S. J. Lee, S. Grossman-Clake, J. Zehnder, B. Hedquist, and E. Comparri, "Determinants of Change in the Regional Urban Heat Island in Metropolitan Phoenix (Arizona, USA) between 1990 and 2004," *Climate Research* 33 (2007): 171–82.

3. R. S. Cerveny and R. C. Balling, "Identification of Anthropogenic Weekly Cycles in Northwest Atlantic Pollution, Precipitation and Tropical Cyclones," *Nature* 394 (1997): 561–62.

4. G. Possehl, *The Indus Civilization: A Contemporary Perspective* (Lanham: Altamira Press, Lanham, 2002).

5. V. N. Misra, "Prehistoric Human Colonization of India," *Journal of Biosciences* 26 (2001): 491–531.

6. G. Singh, "The Indus Valley Culture Seen in the Context of Postglacial Climate and Ecological Studies in North-West India," *Archaeology and Physical Anthropology of Oceania* 6 (1971): 177–89; M. Madella and D. Q. Fuller, "Palaeoecology and the Harappan Civilisation of South Asia: A Reconsideration," *Quaternary Science Reviews* 25 (2006): 1283–1301.

7. Singh, "The Indus Valley Culture Seen in the Context of Postglacial Climate and Ecological Studies in North-West India."

8. R. A. Bryson and T. J. Murray, *Climates of Hunger: Mankind and the World's Changing Weather* (Madison: University of Wisconsin Press, 1977).

9. G. L. Possehl, *Indus Age: The Beginnings* (Philadelphia: University of Pennsylvania Press, 1999); Possehl, *The Indus Civilization.*

10. Ibid., p. 268.

11. G. Singh, R. J. Wasson, and D. P. Agrawal, "Vegetational and Seasonal Climatic Changes since the Last Full Glacial in the Thar Desert, Northwestern India," *Review of Palaeobotany and Palynology* 64 (1990): 351–58.

12. Ibid.

13. M. Staubwasser and H. Weiss, "Holocene Climate and Cultural Evolution in Late Prehistoric–Early Historic West Asia," *Quaternary Research* 66 (2006): 372–87.

14. H. H. Lamb, *Weather, Climate and Human Affairs* (London: Routledge, 1988), p. 212.

15. R. Sinha, R. Kumar, S. Sinha, S. K. Tandon, and M. R. Gibling, "Late Cenozoic Fluvial Successions in Northern and Western India: An Overview and Synthesis," *Quaternary Science Reviews* 26 (2007): 2801–22.

16. Ibid.

17. Bryson and Murray, *Climates of Hunger.*

18. H. Weiss and R. S. Bradley, "Archaeology—What Drives Societal Collapse?" *Science* 291 (2001): 609–10.

19. M. Madella and D. Q. Fuller, "Paleoecology and the Harappan Civilisation of South Asia: A Reconsideration," *Quaternary Science Reviews* 25 (2006): 1283–1301.

20. Abd al-Latif, *The Eastern Key*, translated into English by Kamal Hafuth Zand and John A. and Ivy E. Videan (London: Allen and Unwin, 1965).

CHAPTER 6: THE MYSTERY OF THE EXODUS

1. Psalms 135:7.

2. Exodus 13:32.

3. Exodus 13:17–18.

4. E. D. Oren, ed., *The Sea Peoples and Their World: A Reassessment* (Philadelphia: University Museum, University of Pennsylvania, 2000).

5. Exodus 13:18.

6. Exodus 13:21.

7. 2 Kings 2:11.

8. R. L. Ives, "Behavior of Dust Devils," *Bulletin of the American Meteorological Society* 28 (1947): 168–74.

9. R. Rollo, *On Hail* (London: Edward Stanford, 1893), pp. 75–76.

10. Exodus 13:21.

11. J. E. Weems, *The Tornado* (Garden City, NY: Doubleday & Company, 1977).

12. Exodus 13:19–20.

13. Exodus 14:21–23, 26–29.

14. H. Goedicke, "The Chronology of the Thera/Santorin Explosion," *Zeitschrift für Ägyptische Archäologie Nachbarn* 3 (1992): 57–62.

15. D. Nof and N. Paldor, "Are There Oceanographic Explanations for the Israelites' Crossing of the Red Sea?" *Bulletin of the American Meteorological Society* 73 (1992): 305–14.

16. Ibid.

17. M. Dayan, "The Dividing of the Red Sea according to Natural Sciences" (in Hebrew with English abstract), *Beit Mikra* (publication of the Israel Society for Biblical Research) 2 (1978): 162–77.

18. D. Nof and N. Paldor, "Statistics of Wind over the Red Sea with Application to the Exodus Question," *Journal of Applied Meteorology* 33 (1994): 1017–25.

19. Herodotus, *The Egypt of Herodotus, Being the Second Book, Entitled Euterpe, of the History, on the English version of the Late Prof. George Rawlinson* (London: M. Hopkinson and Company, Ltd., 1924), p. 6.

20. Strabo, *Geography of Strabo*, 3 vols. (London: Henry Bohn, 1854), 3: 188.

21. Defense Mapping Agency, *Sailing Direction [Enroute] Red Sea and the Persian Gulf*, 6th ed., DMA Stock NO. SDPUB172 (Bethesda, MD: Defense Mapping Agency, 1993).

22. R. S. Cerveny and J. A. Shaffer, "Long-Term Equilibrium Tides," *Journal of Geophysical Research-Oceans* 103 (2004): 18801–804.

23. "Trent Tidal Bore," *London Times*, September 16, 1935, p. 9.

24. Deuteronomy 16:1.

25. Exodus 13:14.

26. Nehemiah 2:1.

27. Nof and Paldor, "Are There Oceanographic Explanations for the Israelites' Crossing of the Red Sea?" p. 313.

CHAPTER 7: THE MYSTERIES OF WEATHER AND CLIMATE IN ANCIENT GREECE

1. J. R. March, *Cassell's Dictionary of Classical Mythology* (New York: Cassell, 1998), p. 32.

2. J. C. Kraft, G. Rapp, I. Kayan, and J. V. Luce, "Harbor Areas at Ancient Troy: Sedimentology and Geomorphology Complement Homer's *Iliad*," *Geology* 31 (2003): 163–66.

3. R. S. Cerveny, "Meteorological Assessment of Homer's *Odyssey*," *Bulletin of the American Meteorological Society* 74 (1993): 1025–34.

4. Homer, *Odyssey* 3.158.

5. Ibid., 3.163.

6. Ibid., 9.39–40.

7. Ibid., 3.169–74.
8. Ibid., 3.177.
9. Ibid., 3.183.
10. Ibid., 3.287.
11. Ibid., 3.289–90.
12. Ibid., 4.514.
13. Ibid., 4.500.
14. Ibid., 9.67–69.
15. Ibid., 9.69–71.
16. R. Carpenter, *Discontinuity in Greek Civilization* (Cambridge: University of Cambridge Press, 1966).
17. R. A. Bryson, H. H. Lamb, and D. L. Donley, "Drought and Decline in Mycenae," *Antiquity* 48 (1974): 46–50.
18. Ibid., p. 50.
19. Ibid.

CHAPTER 8: THE MYSTERY OF PETRA, THE ROSE CITY

1. J. Caesar, *De Bello Gallico, Commentaries of Caesar on the Gallic War: The Original Text Reduced to the Natural English Order, with a Literal Interlinear Translation of the First Seven Books* (New York: D. McKay, translated 1964), book 5.
2. J. Taylor, *Petra* (London: Aurum Press, 2005).
3. H. Glueck, *Deities and Dolphins* (New York: Farrar, Straus and Giroux, 1965).
4. L.-A. Bedal, "Desert Oasis: Water Consumption and Display in the Nabataean Capital," *Near Eastern Archaeology* 65 (2002): 225–34; C. R. Ortloff, "The Water Supply and Distribution System of the Nabataean City of Petra (Jordan), 300 BC–AD 300," *Cambridge Archeological Journal* 15 (2005): 93–109; I. Rubin, ed., *The Petra Siq: Nabataean Hydrology Uncovered* (Oakville, CT: David Brown Book Co., 2003).
5. K. W. Russell, "The Earthquake of May 19, A.D. 363," *Bulletin of the American Schools of Oriental Research* 238 (1980): 47–64.
6. T. R. Paradise, "Sandstone Weathering and Aspect in Petra, Jordan," *Zeitschrift Fur Geomorphologie* 46 (2002): 1–17.
7. R. A. Heinlein, *Time Enough for Love* (New York: G. P. Putnam's Sons, 1973), p. 245.

CHAPTER 9: THE MYSTERY OF THE DISAPPEARING SUN

1. T. Short, *A General Chronological History of the Air, Weather, Seasons, Meteors, etc. in Sundary Places and Different Times . . .*, vol. 1 (London: T. Longman and A. Millar, 1749), pp. 64–65.

2. H. H. Lamb, *Climate, History and the Modern World* (London: Routledge, 1982), p. 162.

3. R. B. Strothers, "Mystery Cloud of AD 536," *Nature* 307 (1984): 344–45.

4. Ibid.

5. Ibid.

6. Ibid.

7. T. L. Phipson, *Familiar Letters on Some Mysteries of Nature and Discoveries in Science* (London: Marston, Seale & Rivington, 1876), p. 35.

8. L. F. Kaemtz, *A Complete Course of Meteorology*, translated by C. V. Walker (London: Hippolyte Baillière, 1845), pp. 469–70.

9. H. C. Fritts, *Tree Rings and Climate* (London: Academic Press, 1976).

10. E. Schulman, "Bristlecone Pine, Oldest Known Living Thing," *National Geographic* 113 (1958): 355–72.

11. M. K. Hughes, P. M. Kelly, J. R. Pilcher, and V. C. LaMarche Jr., *Climate from Tree Rings* (Cambridge: Cambridge University Press, 1982).

12. M. G. L. Baillie, *Tree-Ring Dating and Archeology* (Chicago: University of Chicago Press, 1982); M. G. L. Baillie, *Exodus to Arthur: Catastrophic Encounters with Comets* (London: B. T. Batsford Ltd., 1999); M. G. L. Baillie, "The Case for Significant Numbers of Extraterrestrial Impacts through the Late Holocene," *Journal of Quaternary Science* 22 (2007): 101–109.

13. Strothers, "Mystery Cloud of AD 536."

14. G. A. Zielinski, P. A. Mayewski, L. D. Meeker, S. Whitlow, M. S. Twickler, M. Morrison, D. A. Meese, A. J. Gow, and R. B. Alley, "Record of Volcanism since 7000 B.C. from the GISP2 Greenland Ice Core and Implications for the Volcano-Climate System," *Science* 264 (1994): 948–52.

15. Baillie, *Tree-Ring Dating and Archeology*.

16. D. Steel, "Planetary Science: Tunguska at 100," *Nature* 453 (2008): 1157–59.

17. M. G. L. Baillie, "Proposed Re-dating of the European Ice Core Chronology by Seven Years prior to the 7th Century AD," *Geophysical Research Letters* 35 (2008): doi:10.1029/2008GL15813.

18. L. B. Larson, B. M. Vinter, K. R. Briffa, T. M. Melvin, H. B. Clausen, P. D. Jones, M.-L. Siggaard-Andersen, C. U. Hammer, E. Eronen,

H. Grudd, B. E. Gunnarson, R. M. Hantemirov, M. M. Naurzbaev, and K. Nicolussi, "New Ice Core Evidence for a Volcanic Cause of the A.D. 536 Dust Veil," *Geophysical Research Letters* 35 (2008): doi:10.1029/2008 GL04708.

19. Ibid., p. 1.

20. N. Webster, *A Brief History of Epidemic and Pestilential Diseases*, 2 vols. (New York: Burt Franklin, 1799, reprinted 1970), 1: 94.

CHAPTER 10: THE MYSTERY OF THE MAYAN MEGADROUGHT

1. G. H. Haug, D. Gunther, L. C. Peterson, D. M. Sigman, K. A. Hughen, and B. Aeschlimann, "Climate and the Collapse of Maya Civilization," *Science* 299 (2003): 1731–35.

2. H.C. Fritts, *Tree Rings and Climate* (London: Academic Press, 1976).

3. Ibid.

4. R. Acuna-Soto, D. W. Stahle, M. K. Cleaveland, and M. D. Therrell, "Megadrought and Megadeath in 16th Century Mexico," *Emerging Infectious Diseases* 8 (2002): 360–62.

5. Haug, Gunther, Peterson, Sigman, Hughen, and Aeschlimann, "Climate and the Collapse of Maya Civilization."

6. Acuna-Soto, Stahle, Cleaveland, and Therrell, "Megadrought and Megadeath in 16th Century Mexico."

7. T. L. Yates, J. N. Mills, C. A. Parmenter, T. G. Ksiazek, R. R. Parmenter, J. R. Vande Castle, C. H. Calisher, S. T. Nichol, K. D. Abbott, J. C. Young, M. L. Morrison, B. J. Beaty, J. L. Dunnum, R. J. Baker, J. Salazar-Bravo, and C. J. Peters, "The Ecology and Evolutionary History of an Emergent Disease: Hantavirus Pulmonary Syndrome," *Bioscience* 52 (2002): 989–98.

8. R. Acuna-Soto, D. W. Stahle, M. D. Therrell, S. Comez Chavez, and M. K. Cleaveland, "Drought, Epidemic Disease, and the Fall of Classic Period Cultures in Mesoamerica (AD 750–950). Hemorrhagic Fevers as a Cause of Massive Population Loss," *Medical Hypotheses* 65 (2005): 405–409.

9. Ibid., p. 409.

CHAPTER 11: THE MYSTERY OF COLUMBUS'S MISSING HURRICANE

1. A. C. Doyle, *The Complete Sherlock Holmes* (New York: Garden City Publishing Co., 1938), p. 397.

2. D. M. Ludlum, *Early American Hurricanes (1492–1870)* (Boston: American Meteorological Society, 1963), p. 1.

3. R. S. Cerveny and J. S. Hobgood, "Meteorological Implications of the First Voyage of Christopher Columbus," *Bulletin of the American Meteorological Society* 73 (1992): 173–78.

4. R. H. Fuson, trans., *The Log of Christopher Columbus* (Camden, MA: International Marine Publishing, 1992).

5. R. A. Goldsmith and P. L. Richardson, *Numerical Simulations of Columbus' Atlantic Crossings*, Technical Report WHOI-92-14 (Woods Hole, MA: Woods Hole Oceanographic Institution, 1992).

CHAPTER 12: THE MYSTERY OF THE LITTLE ICE AGE'S LOST SUNSPOTS

1. "Historic Natural Events," *Nature* 126 (1930): 976.

2. Ibid., p. 864.

3. J. Evelyn quoted in D. P. Thomson, *Introduction to Meteorology* (Edinburgh and London: W. Blackwood and Sons, 1849), p. 68.

4. E. F. King, ed., *Ten Thousand Wonderful Things* (London: Routledge and Sons, London, 1860; reprinted Gale Research Company, Detroit, 1970), pp. 67–68.

5. "The Thames," *London Times*, February 2, 1814, p. 3.

6. E. R. Floyd, *Great Southern Mysteries (Two Volumes in One)* (New York: Barnes & Noble Books, 1990), 1: 37–40.

7. T. Short, *A General Chronological History of the Air, Weather, Seasons, Meteors, etc. in Sundary Places and Different Times . . .*, vol. 1 (London: T. Longman and A. Millar, 1749), p. 334.

8. J. A. Eddy, "The Maunder Minimum," *Science* 192 (1976): 1189–1202.

9. J. A. Eddy, "The Case of the Missing Sunspots," *Scientific American* 236 (1977): 80.

10. S. Cassini, "An Intimation of Divers Philosophical Particulars, Now Undertaken and Consider'd by Several Ingenious and Learned Men; Here Inserted to Excite Others to Joyn with Them in the Same or the Like Attempts and Observations," *Philosophical Transactions (1665–1678)* 6 (1671): 2216.

11. "New Observations of Spots in the Sun; Made at the Royal Academy of Paris, the 11, 12 and 13th of August 1671," *Philosophical Transactions* 6 (1671): 2253.

12. Eddy, "The Maunder Minimum."

13. Ibid., p. 1193.

14. Ibid., p. 1198.

15. Ibid.

16. Ibid., p. 1189.

17. Q.-S. Ge, J.-Y. Zheng, Z.-X. Hao, P.-Y. Zhang, and W.-C. Wang, "Reconstruction of Historical Climate in China, High-Resolution Precipitation Data from Qing Dynasty Archives," *Bulletin of the American Meteorological Society* 86 (2005): 671–79.

18. D. T. Shindell, G. A. Schmidt, M. E. Mann, D. Rind, and A. Waple, "Solar Forcing of Regional Climate Change during the Maunder Minimum," *Science* 294 (2001): 2149–52.

19. M. E. Mann, R. S. Bradley, and M. K. Hughes, "Global-Scale Temperature Patterns and Climate Forcing over the Past Six Centuries," *Nature* 392 (1998): 779–87; S. McIntyre and R. McKitrick, "Hockey Sticks, Principal Components, and Spurious Significance," *Geophysical Research Letters* 32 (2005): doi:10.1029/2004GL021750.

CHAPTER 13: THE MYSTERY OF THE COMPLACENT EMPIRE'S WET WEATHER

1. Q.-S. Ge, J.-Y. Zheng, Z.-X. Hao, P.-Y. Zhang, and W.-C. Wang, "Reconstruction of Historical Climate in China: High-Resolution Precipitation Data from Qing Dynasty Archives," *Bulletin of the American Meteorological Society* 86 (2005): 671–79.

2. Ibid.

3. Ibid., pp. 674–75.

4. Ibid., p. 675.

5. Ibid.

6. Ibid., pp. 675–76.

7. Q.-S. Ge, J.-Y. Zheng, Y.-Y. Tian, W.X. Wu, X.-Q. Fang, and W.-C. Wang, "Coherence of Climate Reconstruction from Historic Documents in China by Different Studies," *International Journal of Climatology* 28 (2008): 1007–24; J. Jiang, D. Zhang, and K. Fraedrich, "Historic Climate Variability of Wetness in East China (960–1992): A Wavelet Analysis," *International Journal of Climatology* 17 (1997): 969–81.

8. C. C. Huang, J. Pang, X. Zha, H. Su, Y. Jia, and Y. Zhu, "Impact of Monsoonal Climatic Change on Holocene Overbank Flooding along

Sushui River, Middle Reach of the Yellow River, China," *Quaternary Science Reviews* 26 (2007): 2247–64.

9. Jiang, Zhang, and Fraedrich, "Historic Climate Variability of Wetness in East China (960–1992)."

10. W. Qian, Z. Yu, and Y. Zhu, "Spatial and Temporal Variability of Precipitation in East China from 1880 to 1999," *Climate Research* 32 (2006): 209–18.

11. L. Davis, *Natural Disasters (From the Black Plague to the Eruption of Mt. Pinatubo)* (New York: Facts On File, 1992), pp. 72–73.

12. "Human Flesh as Food and Medicine," *New York Times*, July 6, 1878, p. 2.

13. J. R. Nash, *Darkest Hours* (Chicago: Nelson Hall, 1976) p. 115; R. Hewitt, *From Earthquake, Fire and Flood* (New York: Scribner's Sons, 1957), p. 171.

14. Nash, *Darkest Hours*, p. 115.

15. Hewitt, *From Earthquake, Fire and Flood*, p. 171.

16. S. J. Spignesi, *The Odd Index, the Ultimate Compendium of Bizarre and Unusual Facts* (New York: Plume Books, 1994), p. 290.

17. "Historic Natural Events," *Nature* 126 (1930): 153.

18. Spignesi, *The Odd Index*, p. 291.

19. D. Phillips, M. Parfit, and S. Chishom, *Blame It on the Weather: Amazing Weather Facts* (San Diego: Portable Press, San Diego, 1998), p. 68.

20. Spignesi, *The Odd Index*, p. 290.

21. R. Smith, *Catastrophes and Disasters* (Edinburgh: W & R Chambers, 1992), p. 106.

22. Spignesi, *The Odd Index*, p. 290.

CHAPTER 14: THE MYSTERY OF THE PACIFIC HOT TUB

1. C. Darwin, *Journal of Researches into the Natural History and Geology of the Countries Visited during the Voyage of the H.M.S.* Beagle *round the World, under the Command of Capt. Fitz Roy, R.N.*, 2nd. ed. (New York: D. Appleton and Company, 1897), pp. 157–58.

2. H. F. Diaz and V. Markgraf, *El Niño: Historical and Paleoclimatic Aspects of the Southern Oscillation* (New York: Cambridge University Press, 1992).

3. G. T. Walker and E. W. Bliss, "World Weather V," *Memoirs of the Royal Meteorological Society* 4 (1932): 53–84.

4. J. Bjerknes, "Atmospheric Teleconnections from the Equatorial Pacific," *Monthly Weather Review* 97 (1969): 163–72.

5. Ibid., p. 167.
6. Ibid., pp. 167–68.
7. Ibid., pp. 168–70.
8. Ibid., p. 170.
9. K. Wyrtki, "El Niño—The Dynamic Response of the Equatorial Pacific Ocean to Atmospheric Forcing," *Journal of Physical Oceanography* 5 (1975): 572–84; K. Wyrtki, "Sea Level during the 1972 El Niño," *Journal of Physical Oceanography* 7 (1977): 779–87; K. Wyrtki, "The Response of Sea Surface Topography to the 1976 El Niño," *Journal of Physical Oceanography* 9 (1979): 1223–31; K. Wyrtki, "Sea Level Fluctuations in the Pacific during the 1982–83 El Niño," *Geophysical Research Letters* 12 (1985): 125–28.
10. G. T. Mitchum and W. B. White, "Improved Determination of Global Mean Sea Level Variations Using TOPEX/POSEIDON Altimeter Data," *Geophysical Research Letters* 24 (1997): 1331–34.
11. H. F. Diaz and V. Markgraf, eds., *El Niño and the Southern Oscillation: Multiscale Variability and Global and Regional Impacts* (New York: Cambridge University Press, 2000).
12. S. A. Changnon, ed., *El Niño, 1997–1998: The Climate Event of the Century* (New York: Oxford University Press, 2000).
13. R. S. Cerveny and J. A. Shaffer, "The Moon and El Niño," *Geophysical Research Letters* 28 (2001): 25–28.

CHAPTER 15: THE MYSTERY OF TIBET'S WEATHER SPIES

1. T. G. Montgomerie, "On the Geographical Position of Yarkund, and Some Other Places in Central Asia," *Journal of the Royal Geographical Society of London* 36 (1866): 157–72; T. G. Montgomerie, "Report of a Route-Survey Made by Pundit, from Nepal to Lhasa, and Thence through the Upper Valley of the Brahmaputra to Its Source," *Journal of the Royal Geographical Society of London* 38 (1868): 129–219.
2. R. Kipling, *Kim* (London: Macmillan and Co, 1901).
3. H. Pottinger, *Travels in Beloochistan and Sinde* (London: Oxford University Press, 2002), p. 126.
4. Ibid., p. 127.
5. Ibid., p. 129.
6. Ibid.
7. Ibid., pp. 129–30.
8. Ibid., pp. 142–43.
9. Ibid., p. 143.
10. Ibid.

11. Ibid., p. 177.

12. Ibid.

13. Ibid.

14. V. A. Perovsky, *A Narrative of the Russian Military Expedition to Khiva, under General Perofsky*, translated from the Russian for the Foreign Department of the Government of India (Calcutta: Office of Superintendent Government Printing, 1867), p. 69.

15. Ibid., p. 141.

16. Ibid., p. 154.

17. Ibid., p. 159.

18. Ibid., p. 141

19. Ibid., p. 143.

20. Ibid., p. 158.

21. Montgomerie, "On the Geographical Position of Yarkund," p. 157.

22. Ibid., p. 158.

23. Ibid., p. 159.

24. Montgomerie, "Report of a Route-Survey Made by Pundit, from Nepal to Lhasa," p. 153.

25. Montgomerie, "On the Geographical Position of Yarkund," p. 161.

26. Ibid., p. 162.

27. Montgomerie, "Report of a Route-Survey Made by Pundit, from Nepal to Lhasa," p. 141.

28. Ibid., p. 152.

29. Ibid., p. 129.

30. M. G. Marcus and A. J. Brazel, "Environmental Effects on Radiation Fluxes during the Pre-Monsoon," *Mountain Research and Development* 16 (1996): 221–34.

CHAPTER 16: THE MYSTERY OF THE GREAT AMERICAN DUST BOWL

1. M. P. Lawson and C. W. Stockton, "Desert Myth and Climatic Reality," *Annals of the Association of the American Geographers* 71 (1981): 527–35.

2. N. Webster quoted in S. Forry, "Researches in Elucidation of the Distribution of Heat over the Globe . . . ," *American Journal of Science* 47 (1844): 30.

3. Ibid.

4. J. Whitford, "Has the Pacific Railroad Changed the Climate of the Plains," *Scientific American* 21 (1869): 214.

5. E. Gale quoted in C. B. McIntosh, "Use and Abuse of the Timber

Culture Act," *Annals of the Association of American Geographers* 65 (1975): 348.

6. McIntosh, "Use and Abuse of the Timber Culture Act," pp. 347–62.

7. A. R. Greene quoted in McIntosh, "Use and Abuse of the Timber Culture Act," p. 355.

8. H. H. Clayton, *World Weather, Including a Discussion of the Influence of Solar Radiation on the Weather* (New York: Macmillan, 1923).

9. C. W. Stockton and D. M. Meko, "Drought Recurrence in the Great Plains as Reconstructed from Long-Term Tree-Ring Records," *Journal of Climate and Applied Meteorology* 22 (1983): 17–29.

10. J. M. Mitchell, C. W. Stockton, and D. M. Meko, "Drought Cycles in United States and Their Relation to Sunspot Cycles since 1700 AD," *Transactions of the American Geophysical Union* 58 (1977): 694.

11. R. G. Currie, "Variance Contribution of Luni-Solar (M(n)) and Solar Cycle (S-c) Signals to Climate Data," *International Journal of Climatology* 16 (1996): 1343–64.

12. R. G. Currie and D. P. O'Brien, "Deterministic Signals in USA Precipitation," *International Journal of Climatology* 10 (1990): 795–818.

13. R. G. Currie and D. P. O'Brien, "Morphology of Bistable 180-Degrees Phase Switches in 18.6-Year Induced Rainfall over the Northwestern United States of America," *International Journal of Climatology* 9 (1989): 501–25.

14. J. M. Mitchell, C. W. Stockton, and D. M. Meko, "Evidence of a 22-Year Rhythm of Drought in the Western United States Related to the Hale Solar Cycle since the 17th Century," in *Solar Terrestrial Influence on Weather and Climate*, ed. B. M. McCormac and T. A. Seliga (Dordrecht: D. Reidel, 1979), pp. 125–43.

15. E. R. Cook, D. M. Meko, and C. W. Stockton, "A New Assessment of Possible Solar and Lunar Forcing of the Bidecadal Drought Rhythm in the Western United States," *Journal of Climate* 10 (1997): 1343–56.

16. N. J. Mantua, S. R. Hare, Y. Zhang, J. M. Wallace, and R. C. Francis, "A Pacific Interdecadal Climate Oscillation with Impacts on Salmon Production," *Bulletin of the American Meteorological Society* 78 (1997): 1069–79.

CHAPTER 17: THE MYSTERY OF THE CRASHING AIRPLANE AND THE A-BOMB

1. T. T. Fujita, *Proposed Characterization of Tornadoes and Hurricanes by Area and Intensity*, SMRP Res. Paper No. 91 (Chicago: University of Chicago, 1971).

2. University of Chicago News Office, "Tetsuya 'Ted' Fujita, 1920–1998," press release, November 20, 1998, http://www-news.uchicago.edu/releases/98/981120.fujita.shtml (accessed July 28, 2007).

3. Ibid.

4. Ibid.

5. Ibid.

6. T. T. Fujita, *The Downburst, Microburst, and Macroburst*, SMRP Res. Paper No. 210 (Chicago: University of Chicago, 1985), p. 51.

7. Ibid.

CHAPTER 18: THE MYSTERY OF "ELIMINATING THE IMPOSSIBLE"

1. R. A. Bryson, "'All Other Factors Being Constant . . .' A Reconciliation of Several Theories of Climate Change," *Weatherwise* 21 (1968): 56–61.

2. A. Brazel, P. Gober, S. J. Lee, S. Grossman-Clake, J. Zehnder, B. Hedquist, and E. Comparri, "Determinants of Change in the Regional Urban Heat Island in Metropolitan Phoenix (Arizona, USA) between 1990 and 2004," *Climate Research* 33 (2007): 171–82.

3. N. B. Guttman, "Statistical Characteristics of US Historical Climatology Network Temperature Distributions," *Climate Research* 6 (1996): 33–43.

4. R. S. Vose, C. N. Williams Jr., T. C. Peterson, T. R. Karl, and D. R. Easterling, "An Evaluation of the Time of Observation Bias Adjustment in the U.S. Historical Climatology Network," *Geophysical Research Letters* 30 (2003): doi:10.1029/2003GL018111.

5. R. S. Cerveny, B. M. Svoma, R. C. Balling, and R. S. Vose, "The Gregorian Calendar Bias in Monthly Temperature Databases," *Geophysical Research Letters* 35 (2008): doi:10.1029/2008GL035209.

CHAPTER 19: THE MYSTERY OF THE DEVIL IN THE DEEP BLUE SEA

1. B. Franklin, "A Letter from Dr. Benjamin Franklin, to Mr. Alphonsus le Roy, Member of Several Academies, at Paris. Containing Sundry Maritime Observations," *Transactions of the American Philosophical Society* 2 (1786): 294–329.

2. Ibid.

3. W. S. Broecker, "Massive Iceberg Discharges as Triggers for Global

Climate Change," *Nature* 360 (2002): 245–59; W. S. Broecker, "Unpleasant Surprises in the Greenhouse?" *Nature* 328 (1987): 123–26.

4. H. Heinrich, "Origin and Consequences of Cyclic Ice Rafting in the Northeast Atlantic Ocean during the Past 130,000 Years," *Quaternary Research* 29 (1988): 142–52.

5. R. B. Alley, "The Younger Dryas Cold Interval as Viewed from Central Greenland," *Quaternary Science Reviews* 19 (2000): 223.

6. R. B. Alley, "Ice-Core Evidence of Abrupt Climate Change," *Proceedings of the National Academy of Sciences* 97 (2000): 1331–34.

CHAPTER 20: THE MYSTERY OF CLIMATE, COCAINE, AND THE STORYTELLER

1. R. A. Heinlein, *Time Enough for Love* (New York: G. P. Putnam's Sons, 1973), p. 240.

2. Ibid., p. 348.

3. M. McGlade, R. S. Cerveny, and R. Henkel, "Climate and Cocaine," *Nature* 361 (1993): 25.

CHAPTER 21: THE MYSTERY OF THE FUTURE

1. M. Milankovitch, "Canon of Insolation and the Ice Age Problem" (in Yugoslavian), *K. Serb. Acad. Beorg.*, Special Publication 132, English translation by Israel Program for Scientific Translation, Jerusalem, 1969.

2. J. A. Adémar, *Révolutions de la mer (Revolutions of the Sun)* (Paris: Carilian-Goeury et V. Dalmont, 1842).

3. A. Berger, "Milankovitch Theory and Climate," *Reviews of Geophysics* 26 (1988): 624–57.

4. Milankovitch, "Canon of Insolation and the Ice Age Problem."

5. Ibid.

6. A. Berger, "Long-Term Variations of Daily Insolation and Quaternary Climatic Changes," *Journal of the Atmospheric Sciences* 35 (1978): 2362–67.

7. J. D. Hays, J. Imbrie, and N. J. Shackleton, "Variations in the Earth's Orbit: Pacemaker of the Ice Ages," *Science* 194 (1976): 1121–32.

8. Ibid.

9. A. Berger, J. Imbrie, J. Hays, G. Kukla, and B. Saltzman, eds., *Milankovitch and Climate: Understanding the Response to Astronomical Forcing*, 2 vols., NATO ASI Series C (Dordrecht: D. Reidel Publishing, 1982).

10. I. J. Winograd, J. M. Landwehr, K. R. Ludwig, T. B. Coplen, and A. C. Riggs, "Duration and Structure of the Past Four Interglaciations," *Quaternary Research* 48 (1997): 141–54.

11. T. B. Coplen, I. J. Winograd, J. M. Landwehr, and A. C. Riggs, "500,000-Year Stable Carbon Isotopic Record from Devil's Hole, Nevada," *Science* 263 (1994): 361–65.

12. J. A. Shaffer, R. S. Cerveny, and R. A. Dorn, "Radiation Windows as Indicators of an Astronomical Influence on the Devil's Hole Chronology," *Geology* 24 (1996): 1017–20.

13. L. F. Richardson, *Weather Prediction by Numerical Process* (New York: Dover Publications, Inc., 1965).

14. R. P. Turco, O. B. Toon, T. P. Ackerman, J. B. Pollack, and C. Sagan, "Nuclear Winters—Global Consequence of Multiple Nuclear Explosions," *Science* 222 (1983): 1283–92.

15. S. H. Schneider and S. L. Thompson, "Simulating the Climatic Effect of Nuclear-War," *Nature* 333 (1988): 221–27.

16. J. Raisanen, "CO_2-Induced Climate Change in CMIP2 Experiments: Quantification of Agreement and Role of Internal Variability," *Journal of Climate* 14 (2001): 2088–2104.

17. P. Lemesurier, *The Unknown Nostradamus: 500 Anniversary Biography* (London: O Publishing, 2003).

CHAPTER 22: THE SEVEN SOLUTIONS

1. S. H. Schneider, "Earth Systems Engineering and Management," *Nature* 409 (2001): 417–21.

2. Ibid.

ACKNOWLEDGMENTS

1. Robert A. Heinlein, *Time Enough for Love* (New York: G. P. Putnam's Sons, 1973), p. 353.

BIBLIOGRAPHY

Abd al-Latif. *The Eastern Key*. Translated by Kamal Hafuth Zand, and John A. and Ivy E. Videan. London: Allen and Unwin, 1965.

Acuna-Soto, R., D. W. Stahle, M. K. Cleaveland, and M. D. Therrell. "Megadrought and Megadeath in 16th Century Mexico." *Emerging Infectious Diseases* 8 (2002): 360–62.

Acuna-Soto, R., D. W. Stahle, M. D. Therrell, S. Comez Chavez, and M. K. Cleaveland. "Drought, Epidemic Disease, and the Fall of Classic Period Cultures in Mesoamerica (AD 750–950). Hemorrhagic Fevers as a Cause of Massive Population Loss." *Medical Hypotheses* 65 (2005): 405–409.

Adémar, J. A. *Révolutions de la mer (Revolutions of the Sun)*. Paris: Carilian-Goeury et V. Dalmont, 1842.

Agustsdottir, A. M., E. J. Barron, K. L. Bice, L. A. Colarusso, J. L. Cookman, B. A. Cosgrove, J. L. De Lurio, J. F. Frakes, L. A. Frakes, C. J. Moy, T. D. Olszewski, R. D. Pancost, C. J. Poulsen, C. M. Ruffner, D. G. Sheldon, and T. S. White. "Storm Activity in Ancient Climates 2. An Analysis Using Climate Simulations and Sedimentary Structures." *Journal of Geophysical Research—Atmospheres* 104 (1999): 27295–320.

Alley, R. B. "Ice-Core Evidence of Abrupt Climate Change." *Proceedings of the National Academy of Sciences* 97 (2000): 1331–34.

———. "The Younger Dryas Cold Interval as Viewed from Central Greenland." *Quaternary Science Reviews* 19 (2000): 213–26.

Alvarez, L. W., W. Alvarez, F. Asaro, and H. V. Michel. "Extraterrestrial Cause for the Cretaceous-Tertiary Extinction: Experimental Results and Theoretical Interpretation." *Science* 208 (1980): 1095–1108.

Alvarez, W., P. Claeys, and S. W. Kieffer. "Emplacement of Cretaceous-Tertiary Boundary Shocked Quartz from Chicxulub Crater." *Science* 269 (1995): 930–35.

Ambrose, S. H. "Late Pleistocene Human Population Bottlenecks, Volcanic Winter, and Differentiation of Modern Humans." *Journal of Human Evolution* 34 (1998): 623–51.

Baillie, M. G. L. "The Case for Significant Numbers of Extraterrestrial Impacts through the Late Holocene." *Journal of Quaternary Science* 22 (2007): 101–109.

———. *Exodus to Arthur: Catastrophic Encounters with Comets*. London: B. T. Batsford Ltd., 1999.

———. "Proposed Re-dating of the European Ice Core Chronology by Seven Years prior to the 7th Century AD." *Geophysical Research Letters* 35 (2008): doi:10.1029/2008GL15813.

———. *Tree-Ring Dating and Archeology*. Chicago: University of Chicago Press, 1982.

Barron, E. J., and W. M. Washington. "The Role of Geographic Variables in Explaining Paleoclimates: Results from Cretaceous Climate Model Sensitivity Studies." *Journal of Geophysical Research—Atmospheres* 89 (1984): 1267–79.

Beckman, J. E., and T. J. Mahoney. "The Maunder Minimum and Climate Change: Have Historical Records Aided Current Research?" In *Library and Information Services in Astronomy III (LISA III)*, edited by U. Grothkopf, H. Andernach, S. Stevens-Rayburn, and M. Gomez. San Francisco: Astronomical Society of the Pacific, 1998.

Bedal, L.-A. "Desert Oasis: Water Consumption and Display in the Nabataean Capital." *Near Eastern Archaeology* 65 (2002): 225–34.

Bell, T. L., D. Rosenfeld, K. M. Kim, J.-M. Yoo, M. I. Lee, and M. Hahnenberger. "Midweek Increase in US Summer Rain and Storm Heights Suggests Air Pollution Invigorates Rainstorms." *Journal of Geophysical Research—Atmospheres* 113 (D2) (2008): D02209.

Berger, A. "Long-Term Variations of Daily Insolation and Quaternary Climatic Changes." *Journal of the Atmospheric Sciences* 35 (1978): 2362–67.

———. "Milankovitch Theory and Climate." *Reviews of Geophysics* 26 (1988): 624–57.

———. "Support for Astronomical Theory of Climatic Change." *Nature* 269 (1977): 44–45.

Berger, A., J. Imbrie, J. Hays, G. Kukla, and B. Saltzman, eds. *Milankovitch and Climate: Understanding the Response to Astronomical Forcing.* 2 vols. NATO ASI Series C. Dordrecht: D. Reidel Publishing, 1982.

Bjerknes, J. "Atmospheric Teleconnections from the Equatorial Pacific." *Monthly Weather Review* 97 (1969): 163–72.

Bouregeois, J., T. A. Hansen, P. L. Wiberg, and E. G. Kauffman. "A Tsunami Deposit at the Cretaceous-Tertiary Boundary in Texas." *Science* 241 (1988): 567–70.

Bradley, R. S., M. K. Hughes, and H. F. Diaz. "Climate in Medieval Times." *Science* 302 (2003): 404–405.

Brazel, A., P. Gober, S. J. Lee, S. Grossman-Clake, J. Zehnder, B. Hedquist, and E. Comparri. "Determinants of Change in the Regional Urban Heat Island in Metropolitan Phoenix (Arizona, USA) between 1990 and 2004." *Climate Research* 33 (2007): 171–82.

Bressler, B. O., and W. U. Reimold. "Terrestrial Impact Melt Rock and Glass." *Earth-Science Reviews* 56 (2001): 205–84.

Broecker, W. S. "Massive Iceberg Discharges as Triggers for Global Climate Change." *Nature* 360 (2002): 245–59.

———. "Unpleasant Surprises in the Greenhouse?" *Nature* 328 (1987): 123–26.

Brown, W. R., and N. D. Anderson. *Famines (Historical Catastrophes).* Reading, MA: Addison-Wesley, 1976.

Bryson, R. A. "'All Other Factors Being Constant . . .'" *Weatherwise* 21 (1968): 56–61.

Bryson, R. A., H. H. Lamb, and D. L. Donley. "Drought and Decline in Mycenae." *Antiquity* 48 (1974): 46–50.

Bryson, R. A., and T. J. Murray. *Climates of Hunger: Mankind and the World's Changing Weather.* Madison: University of Wisconsin Press, 1977.

Butzer, K. W. *Environment and Archeology.* Chicago: Aldine Publishing, 1964.

Caesar, J. *De Bello Gallico. Commentaries of Caesar on the Gallic War: The Original Text Reduced to the Natural English Order, with a Literal Interlinear Translation of the First Seven Books* (New York: D. McKay, trans. 1964).

Calder, W. M., III, and D. A. Traill, eds. *Myth, Scandal, and History: The Heinrich Schliemann Controversy and a First Edition of the Mycenaean Diary.* Detroit: Wayne State University Press, 1986.

Carpenter, R. *Discontinuity in Greek Civilization.* Cambridge: University of Cambridge Press, 1966.

Cassini, S. "An Intimation of Divers Philosophical Particulars, Now Undertaken and Consider'd by Several Ingenious and Learned Men; Here Inserted to Excite Others to Joyn with Them in the Same or the Like

Attempts and Observations." *Philosophical Transactions (1665–1678)* 6 (1671): 2216–19.

Cerveny, N. V., R. Kaldenberg, J. Reed, D. S. Whitley, J. Simon, and R. I. Dorn. "A New Strategy for Analyzing the Chronometry of Constructed Rock Features in Deserts." *Geoarchaeology* 21 (2006): 281–303.

Cerveny, R. S. "The Day the Dinosaurs Died." *Weatherwise* 51 (1998): 13–16.

———. "Meteorological Assessment of Homer's *Odyssey*." *Bulletin of the American Meteorological Society* 74 (1993): 1025–34.

———. "The Meteorological Observations of Charles Darwin aboard the H.M.S. *Beagle*." *Bulletin of the American Meteorological Society* 86 (2003): 1295–1301.

———. "Orbital Signals in the Diurnal Cycle of Radiation." *Journal of Geophysical Research—Atmospheres* 96 (1991): 17209–215.

Cerveny, R. S., and R. C. Balling. "Identification of Anthropogenic Weekly Cycles in Northwest Atlantic Pollution, Precipitation and Tropical Cyclones." *Nature* 394 (1997): 561–62.

Cerveny, R. S., and J. S. Hobgood. "The Weather Luck of Christopher Columbus." *Weatherwise* 45 (1992): 34–26.

Cerveny, R. S., and M. Marcus. "Elements of Espionage." *Weatherwise* 47 (1997): 14–21.

Cerveny, R. S., and J. A. Shaffer. "Long-Term Equilibrium Tides." *Journal of Geophysical Research—Oceans* 103 (2004): 18801–804.

———. "The Moon and El Niño." *Geophysical Research Letters* 28 (2001): 25–28.

Cerveny, R. S., B. M. Svoma, R. C. Balling, and R. S. Vose. "The Gregorian Calendar Bias in Monthly Temperature Databases." *Geophysical Research Letters* 35 (2008): doi:10.1029/2008GL035209.

Changnon, S. A., ed. *El Niño, 1997–1998: The Climate Event of the Century*. New York: Oxford University Press, 2000.

Changnon, S. A., F. A. Huff, and R. G. Semonin. "METROMEX: An Investigation of Inadvertent Weather Modification." *Bulletin of the American Meteorological Society* 52 (1971): 958–67.

Clayton, H. H. *World Weather, Including a Discussion of the Influence of Solar Radiation on the Weather*. New York: Macmillan, 1923.

Cook, E. R., D. M. Meko, and C. W. Stockton. "A New Assessment of Possible Solar and Lunar Forcing of the Bidecadal Drought Rhythm in the Western United States." *Journal of Climate* 10 (1997): 1343–56.

Coplen, T. B., I. J. Winograd, J. M. Landwehr, and A. C. Riggs. "500,000-Year Stable Carbon Isotopic Record from Devil's Hole, Nevada." *Science* 263 (1994): 361–65.

Covey, C., S. L. Thompson, P. R. Weissman, and M. C. MacCracken. "Global Climatic Effects of Atmospheric Dust from an Asteroid or

Comet Impact on Earth." *Global and Planetary Change* 9 (1994): 263–73.

Crutzen, P. J. "Mass Extinctions: Acid Rain at the K/T Boundary." *Nature* 330 (1987): 108–109.

Currie, R. G. "Variance Contribution of Luni-Solar (M(n)) and Solar Cycle (S-c) Signals to Climate Data." *International Journal of Climatology* 16 (1996): 1343–64.

Currie, R. G., and D. P. O'Brien. "Deterministic Signals in Precipitation Records from the American Corn Belt." *International Journal of Climatology* 10 (1990): 179–89.

———. "Morphology of Bistable 180-Degrees Phase Switches in 18.6-Year Induced Rainfall over the Northwestern United States of America." *International Journal of Climatology* 9 (1989): 501–25.

Darwin, C. *Journal of Researches into the Natural History and Geology of the Countries Visited during the Voyage of the H.M.S.* Beagle *round the World, under the Command of Capt. FitzRoy, R.N.* 2nd ed. New York: D. Appleton and Company, 1897.

Davis, L. *Natural Disasters (From the Black Plague to the Eruption of Mt. Pinatubo)*. New York: Facts On File, 1992.

Dayan, M. "The Dividing of the Red Sea according to Natural Sciences" (in Hebrew with English abstract). *Beit Mikra* (publication of the Israel Society for Biblical Research) 2 (1978): 162–77.

Defant, A. *Physical Oceanography*. New York: Macmillan, 1961.

Defense Mapping Agency. *Sailing Direction [enroute] Red Sea and the Persian Gulf*. 6th ed. DMA Stock NO. SDPUB172. Bethesda, MD: Defense Mapping Agency, 1993.

Diaz, H. F., and V. Markgraf. *El Niño: Historical and Paleoclimatic Aspects of the Southern Oscillation*. New York: Cambridge University Press, 1992.

Dorn, R. I., D. B. Bamforth, T. A. Cahill, J. C. Dhrenwend, B. D. Turrin, D. J. Donahue, A. J. T. Jull, A. Long, M. E. Macko, E. B. Weil, D. S. Whitley, and T. H. Zabel. "Cation-Ratio and Accelerator Radiocarbon Dating of Rock Vanish on Mojave Artifacts and Landforms." *Science* 231 (1986): 830–33.

Dorn, R. I., P. B. Clarkson, M. F. Nobbs, L. L. Loendorf, and D. S. Whitley. "New Approach to the Radiocarbon Dating of Rock Varnish, with Examples from Drylands." *Annals of the Association of American Geographers* 82 (1992): 136–51.

Doyle, A. C. *The Complete Sherlock Holmes*. New York: Garden City Publishing, 1938.

Eddy, J. A. "The Case of the Missing Sunspots." *Scientific American* 236 (1977): 80–91.

———. "The Maunder Minimum." *Science* 192 (1976): 1189–1202.

EPICA Community Members (56 authors). "Eight Glacial Cycles from an Antarctic Ice Core." *Nature* 429 (2004): 628.

Floyd, E. R. *Great Southern Mysteries (Two Volumes in One)*. New York: Barnes & Noble, 1990.

Forry, S. "Researches in Elucidation of the Distribution of Heat over the Globe . . ." *American Journal of Science* 47 (1844): 30.

Forster, P. M. D., and S. Solomon. "Observations of a 'Weekend Effect' in Diurnal Temperature Range." *Proceedings of the National Academy of Sciences* 100 (2003): 11225–230.

Franklin, B. "A Letter from Dr. Benjamin Franklin, to Mr. Alphonsus le Roy, Member of Several Academies, at Paris. Containing Sundry Maritime Observations." *Transactions of the American Philosophical Society* 2 (1786): 294–329.

Fritts, H. C. *Tree Rings and Climate*. London: Academic Press, 1976.

Fujita, T. T. *The Downburst, Microburst, and Macroburst*. SMRP Res. Paper No. 210 [NTIS No. PB85-148880]. Chicago: University of Chicago, 1985.

———. *Manual of Downburst Identification for Project NIMROD*. SMRP Res. Paper No. 156 [NTIS No. N78-30771/7GI]. Chicago: University of Chicago, 1978.

———. *Spearhead Echo and Downburst near the Approach End of a John F. Kennedy Airport Runway, New York City*. SMRP Res. Paper No. 137 [NTIS No. N76-2184/1GI]. Chicago: University of Chicago, 1976.

Fujita, T. T., and H. R. Byers. "Spearhead Echo and Downbursts in the Crash of an Airliner." *Monthly Weather Review* 105 (1977).

Fujita, T. T., and F. Caracena. "An Analysis of Three Weather-Related Accidents." *Bulletin of the American Meteorological Society* 58 (1977): 1164–81.

Fuson, R. H., trans. *The Log of Christopher Columbus*. Camden, MA: International Marine Publishing, 1992.

Ge, Q.-S., J.-Y. Zheng, Z.-X. Hao, P.-Y. Zhang, and W.-C. Wang. "Reconstruction of Historical Climate in China, High-Resolution Precipitation Data from Qing Dynasty Archives." *Bulletin of the American Meteorological Society* 86 (2005): 671–79.

Ge, Q.-S., J.-Y. Zheng, Y.-Y. Tian, W. X. Wu, X.-Q. Fang, and W.-C. Wang. "Coherence of Climate Reconstruction from Historic Documents in China by Different Studies." *International Journal of Climatology* 28 (2008): 1007–24.

Giberne, A. *The Ocean of Air (Meteorology for Beginners)*. London: Seeley, 1894.

Gill, R. B. *The Great Maya Droughts: Water, Life and Death*. Albuquerque: University of New Mexico Press, 2000.

Glueck, H. *Deities and Dolphins*. New York: Farrar, Straus and Giroux, 1965.

Goedicke, H. "The Chronology of the Thera/Santorin Explosion." *Zeitschrift für Ägyptische Archäologie Nachbarn* 3 (1992): 57–62.

Goldsmith, R. A., and P. L. Richardson. *Numerical Simulations of Columbus' Atlantic Crossings*. Technical Report WHOI-92-14. Woods Hole: Woods Hole Oceanographic Institution, 1992.

Guttman, N. B. "Statistical Characteristics of US Historical Climatology Network Temperature Distributions." *Climate Research* 6 (1996): 33–43.

Harrer, H. *Seven Years in Tibet*. London: Rupert Hart-Davis, 1953.

Hartwig, G. *The Aerial World*. London: Longmans, Green, 1887.

Haug, G. H., D. Gunther, L. C. Peterson, D. M. Sigman, K. A. Hughen, and B. Aeschlimann. "Climate and the Collapse of Maya Civilization." *Science* 299 (2003): 1731–35.

Hawkins, T. W., A. J. Brazel, W. L. Stefanov, W. Bigler, and E. M. Saffell. "The Role of Rural Variability in Urban Heat Island Determination for Phoenix, Arizona." *Journal of Applied Meteorology* 43 (2004): 476–86.

Heinlein, Robert A. *Time Enough for Love*. New York: G. P. Putnam's Sons, 1973.

Heinrich, H. "Origin and Consequences of Cyclic Ice Rafting in the Northeast Atlantic Ocean during the Past 130,000 Years." *Quaternary Research* 29 (1988): 142–52.

Herodotus. *The Egypt of Herodotus, Being the Second Book, Entitled Euterpe, of the History, on the English Version of the Late Prof. George Rawlinson*. London: M. Hopkinson and Company, 1924.

Hewitt, R. *From Earthquake, Fire and Flood*. New York: Scribner's Sons, 1957.

Hildebrand, A. R., G. T. Penfield, D. A. Kring, M. Pilkington, A. Camargo, S. B. Jacobsen, and W. V. Boynton. "Chicxulub Crater: A Possible Cretaceous/Tertiary Boundary Impact Crater on the Yucatan Peninsula, Mexico." *Geology* 19 (1991): 867–71.

"Historic Natural Events." *Nature* 125 (1930): 112.

"Historic Natural Events." *Nature* 126 (1930): 976.

"History of Excessive Winters." *Graham's Magazine* 42 (1853): 11–14.

Hopkirk, P. *Foreign Devils on the Silk Road*. London: John Murray Publishers, 1980.

———. *The Great Game*. New York: Kodansha International, 1992.

———. *Trespassers on the Roof of the World*. Los Angeles: J. P. Tarcher, 1983.

Hoyt, D. V., and K. H. Schatten. *The Role of the Sun in Climate Change*. New York: Oxford University Press, 1997.

Huang, C. C., J. Pang, X. Zha, H. Su, Y. Jia, and Y. Zhu. "Impact of Monsoonal Climatic Change on Holocene Overbank Flooding along Sushui

River, Middle Reach of the Yellow River, China." *Quaternary Science Reviews* 26 (2007): 2247–64.

Hughes, M. K., P. M. Kelly, J. R. Pilcher, and V. C. LaMarche Jr. *Climate from Tree Rings*. Cambridge: Cambridge University Press, 1982.

"Human Flesh as Food and Medicine." *New York Times*, July 6, 1878, p. 2.

Humphreys, W. J. *Physics of the Air*. New York: McGraw-Hill, 1940.

Huntington, E. *The Character of Races as Influenced by Physical Environment, Natural Selection and Historical Development*. New York: C. Scribner's Sons, 1924.

———. *Civilization and Climate*. New Haven, CT: Yale University Press, 1924.

Ives, R. L. "Behavior of Dust Devils." *Bulletin of the American Meteorological Society* 28 (1947): 168–74.

Jiang, J., D. Zhang, and K. Fraedrich. "Historic Climate Variability of Wetness in East China (960–1992): A Wavelet Analysis." *International Journal of Climatology* 17 (1997): 969–81.

Jones, M. T., R. S. J. Sparks, and P. J. Valdes. "The Climatic Impact of Supervolcanic Ash Blankets." *Climate Dynamics* 29 (2007): 553–64.

Kaemtz, L. F. *A Complete Course of Meteorology*. Translated by C. V. Walker. London: Hippolyte Bailliére, 1845.

King, E. F., ed. *Ten Thousand Wonderful Things*. London: Routlege and Sons, 1860; reprinted, Detroit: Gale Research Company, 1970.

Kipling, R. *Kim*. London: Macmillan, 1901.

Kraft, J. C., G. Rapp, I. Kayan, and J. V. Luce. "Harbor Areas at Ancient Troy: Sedimentology and Geomorphology Complement Homer's *Iliad*." *Geology* 31 (2003): 163–66.

Kring, D. A. "Air Blast Produced by the Meteor Crater Impact and a Reconstruction of the Affected Environment." *Meteoritics and Planetary Science* 32 (1997): 517–30.

Kruge, M. A., B. A. Stankiewicz, J. C. Crelling, A. Montanaria, and D. F. Bensley. "Possible Charcoal in Cretaceous-Tertiary Boundary Strata: Evidence for Catastrophic Firestorm and Megawave." *Geochimica et Cosmochimica Acta* 58 (1994): 1393–97.

Lamb, H. H. *Climate, History and the Modern World*. London: Routledge, 1982.

———. *Climate, History and the Modern World*. 2nd ed. London: Routledge, 1995.

———. *Weather, Climate and Human Affairs*. London: Routledge, 1988.

Lang, C., M. Leuenberger, J. Schwander, and S. Johnsen. "16°C Rapid Temperature Variation in Central Greenland 70,000 Years Ago." *Science* 286 (1999): 934–37.

Larson, L. B., B. M. Vinter, K. R. Briffa, T. M. Melvin, H. B. Clausen, P. D. Jones, M.-L. Siggaard-Andersen, C. U. Hammer, E. Eronen, H. Grudd, B. E. Gunnarson, R. M. Hantemirov, M. M. Naurzbaev, and K. Nicolussi. "New Ice Core Evidence for a Volcanic Cause of the A.D. 536 Dust Veil." *Geophysical Research Letters* 35 (2008): doi: 10.1029/2008GL04708.

Latif, M., and T. P. Barnett. "Causes of Decadal Climate Variability over the North Pacific and North America." *Science* 266 (1994): 634–37.

Lawson, M. P. "A Dendroclimatological Interpretation of the Great American Desert." *Proceedings of the Association of American Geographers* 3 (1971): 109–14.

Lawson, M. P., and C. W. Stockton. "Desert Myth and Climatic Reality." *Annals of the Association of the American Geographers* 71 (1981): 527–35.

Lemesurier, P. *The Unknown Nostradamus: 500 Anniversary Biography*. London: O Publishing, 2003.

Lomax, B., D. Beerling, G. Upchurch, and B. Ott-Bliesner. "Rapid (10-Yr.) Recovery of Terrestrial Productivity in a Simulation Study of the Terminal Cretaceous Impact Event." *Earth and Planetary Science Letters* 192 (2001): 137–44.

Ludlam, D. M. *Early American Hurricanes (1492–1870)*. Boston: American Meteorological Society, 1963.

Madella, M., and D. Q. Fuller. "Paleoecology and the Harappan Civilisation of South Asia: A Reconsideration." *Quaternary Science Reviews* 25 (2006): 1283–1301.

Mandelbaum, A. *The Odyssey of Homer*. Berkeley: University of California Press, 1990.

Mann, M. E., R. S. Bradley, and M. K. Hughes. "Global-Scale Temperature Patterns and Climate Forcing over the Past Six Centuries." *Nature* 392 (1998): 779–87.

Mantua, N. J., S. R. Hare, Y. Zhang, J. M. Wallace, and R. C. Francis. "A Pacific Interdecadal Climate Oscillation with Impacts on Salmon Production." *Bulletin of the American Meteorological Society* 78 (1997): 1069–79.

March, J. R. *Cassell's Dictionary of Classical Mythology*. New York: Cassell, 1998.

Marcus, M. G., and A. J. Brazel. "Environmental Effects on Radiation Fluxes during the Pre-Monsoon." *Mountain Research and Development* 16 (1996): 221–34.

McGlade, M., R. S. Cerveny, and R. Henkel. "Climate and Cocaine." *Nature* 361 (1993): 25–26.

McIntosh, C. B. "Use and Abuse of the Timber Culture Act." *Annals of the Association of American Geographers* 65 (1975): 347–62.

McIntyre, S., and R. McKitrick. "Hockey Sticks, Principal Components, and Spurious Significance." *Geophysical Research Letters* 32 (2005): doi: 10.1029/2004GL021750.

Meloshi, H. J., N. M. Schneider, K. J. Zahnle, and D. Latham. "Ignition of Global Wildfires at the Cretaceous/Tertiary Boundary." *Nature* 343 (1990): 251–54.

Milankovitch, M. "Canon of Insolation and the Ice Age Problem." *K. Serb. Acad. Beorg.* Special Publication 132, English translation by Israel Program for Scientific Translation, Jerusalem, 1969.

Miller, C. F., and D. A. Wark. "Supervolcanoes and Their Explosive Supereruptions." *Elements* 4 (2008): 11–15.

Misra, V. N. "Prehistoric Human Colonization of India." *Journal of Biosciences* 26 (2001): 491–531.

Mitchell, J. M., C. W. Stockton, and D. M. Meko. "Drought Cycles in United States and Their Relation to Sunspot Cycles since 1700 AD." *Transactions of the American Geophysical Union* 58 (1977).

———. "Evidence of a 22-Year Rhythm of Drought in the Western United States Related to the Hale Solar Cycle since the 17th Century." In *Solar Terrestrial Influence on Weather and Climate*, edited by B. M. McCormac and T. A. Seliga. Dordrecht: D. Reidel, 1979.

Mitchum, G. T., and W. B. White. "Improved Determination of Global Mean Sea Level Variations Using TOPEX/POSEIDON Altimeter Data." *Geophysical Research Letters* 24 (1997): 1331–34.

Montgomerie, T. G. "On the Geographical Position of Yarkund, and Some Other Places in Central Asia." *Journal of the Royal Geographical Society of London* 36 (1866): 157–72.

———. "Report of a Route-Survey Made by Pundit, from Nepal to Lhasa, and Thence through the Upper Valley of the Brahmaputra to Its Source." *Journal of the Royal Geographical Society of London* 38 (1868): 129–219.

Mori, F., R. Ponti, A. Messina, M. Flieger, V. Havlicek, and M. Sinibaldi. "Chemical Characterization and AMS Radiocarbon Dating of the Binder of a Prehistoric Rock Pictograph at Tadrart Acacus, Southern West Libya." *Journal of Cultural Heritage* 7 (2006): 344–49.

Morison, S. E. *Admiral of the Ocean Sea: A Life of Christopher Columbus.* New York: Little, Brown, 1942.

Nash, J. R. *Darkest Hours.* Chicago: Nelson Hall, 1976.

"New Observations of Spots in the Sun; Made at the Royal Academy of Paris, the 11, 12 and 13th of August 1671." *Philosophical Transactions* 6 (1671): 2253.

Nof, D., and N. Paldor. "Are There Oceanographic Explanations for the Israelites' Crossing of the Red Sea?" *Bulletin of the American Meteorological Society* 73 (1992): 305–14.

———. "Statistics of Wind over the Red Sea with Application to the Exodus Question." *Journal of Applied Meteorology* 33 (1994): 1017–25.

Oren, E. D., ed. *The Sea Peoples and Their World: A Reassessment.* Philadelphia: University Museum, University of Pennsylvania, 2000.

Ortloff, C. R. "The Water Supply and Distribution System of the Nabataean City of Petra (Jordan), 300 BC–AD 300." *Cambridge Archeological Journal* 15 (2005): 93–109.

Paradise, T. R. "Sandstone Weathering and Aspect in Petra, Jordan." *Zeitschrift Fur Geomorphologie* 46 (2002): 1–17.

Perovsky, Vasily Aleksyeevitch. *A Narrative of the Russian Military Expedition to Khiva, under General Perofsky.* Translated from the Russian for the Foreign Department of the Government of India. Calcutta: Office of Superintendent Government Printing, 1867.

Phillips, D., M. Parfit, and S. Chishom. *Blame It on the Weather: Amazing Weather Facts.* San Diego: Portable Press, 1998.

Phipson, T. L. *Familiar Letters on Some Mysteries of Nature and Discoveries in Science.* London: Marston, Seale & Rivington, 1876.

Possehl, G. L. "Climate and the Eclipse of the Ancient Cities of the Indus." In *Third Millennium BC Climate Change and Old World Collapse*, edited by H. N. Dalfes, G. Kukla, and W. Weiss. NATO ASI Series 1, volume 49. New York: Springer, 1997.

———. *Indus Age. The Beginnings.* Philadelphia: University of Pennsylvannia Press, 1999.

———. *The Indus Civilization: A Contemporary Perspective.* Lanham, MD: AltaMira Press, 2002.

Pottinger, H. *Travels in Beloochistan and Sinde.* London: Oxford University Press, 2002.

Qian, W., Z. Yu, and Y. Zhu. "Spatial and Temporal Variability of Precipitation in East China from 1880 to 1999." *Climate Research* 32 (2006): 209–18.

Raisanen, J. "CO_2-Induced Climate Change in CMIP2 Experiments: Quantification of Agreement and Role of Internal Variability." *Journal of Climate* 14 (2001): 2088–2104.

Rampino, M. R. "Supereruptions as a Threat to Civilizations on Earth-Like Planets." *Icarus* 156 (2002): 562–69.

Rampino, M. R., and S. Self. "Historic Eruptions of Tambora (1815), Krakatau (1883), and Agung (1963), Their Stratospheric Aerosols, and Climatic Impact." *Quaternary Research* 18 (1982): 127–43.

———. "Volcanic Winter and Accelerated Glaciation following the Toba Super-Eruption." *Nature* 359 (1992): 50–52.

Richardson, L. F. *Weather Prediction by Numerical Process.* New York: Dover, 1965.

Rollo, R. *On Hail.* London: Edward Stanford, 1893.

Rowbotham, F. W. *The Severn Bore.* London: MacDonald, 1964.

Rubin, I., ed. *The Petra Siq: Nabataean Hydrology Uncovered.* Oakville, CT: David Brown, 2003.

Runnels, C. *The Archaeology of Heinrich Schliemann: An Annotated Bibliographic Handlist.* Boston: Archaeological Institute of America, 2007.

Russell, K. W. "The Earthquake of May 19, A.D. 363." *Bulletin of the American Schools of Oriental Research* 238 (1980): 47–64.

Santayana, G. *Reason in Common Sense*, vol. 1 of *The Life of Reason: or, The Phases of Human Progress*. New York: C. Scribner's Sons, 1906.

Schneider, S. H. "Earth Systems Engineering and Management." *Nature* 409 (2001): 417–21.

Schneider, S. H., and S. L. Thompson. "Simulating the Climatic Effect of Nuclear-War." *Nature* 333 (1988): 221–27.

Schott, W. *Deep-Sea Sediments of the Indian Ocean*. Tulsa: American Association of Petroleum Geology, 1939.

Schulman, E. "Bristlecone Pine, Oldest Known Living Thing." *National Geographic* 113 (1958): 355–72.

Shaffer, J. A., R. S. Cerveny, and R. A. Dorn. "Radiation Windows as Indicators of an Astronomical Influence on the Devil's Hole Chronology." *Geology* 24 (1996): 1017–20.

Shaw, B. D. "Climate, Environment, and History: The Case of Roman North Africa." In *Climate and History: Studies in Past Climates and Their Impact on Man*, edited by T. M. L. Wigley, M. J. Ingram, and G. Farmer. New York: University of Cambridge Press, 1981.

Shindell, D. T., G. A. Schmidt, M. E. Mann, D. Rind, and A. Waple. "Solar Forcing of Regional Climate Change during the Maunder Minimum." *Science* 294 (2001): 2149–52.

Short, T. *A General Chronological History of the Air, Weather, Seasons, Meteors, etc. in Sundary Places and Different Times*, vol. 1. London: T. Longman and A. Millar, 1749.

Sigurdsson, H., Ph. Bonté, L. Turpin, M. Chaussidon, N. Metrich, M. Steinberg, Ph. Pradel, and S. D'Hondt. "Geochemical Constraints on Source Region of Cretaceous/Tertiary Impact Glasses." *Nature* 353 (1991): 839–42.

Singh, G. "The Indus Valley Culture Seen in the Context of Postglacial Climate and Ecological Studies in North-West India." *Archaeology and Physical Anthropology of Oceania* 6 (1971): 177–89.

Singh, G., R. J. Wasson, and D. P. Agrawal. "Vegetational and Seasonal Climatic Changes since the Last Full Glacial in the Thar Desert, Northwestern India." *Review of Palaeobotany and Palynology* 64 (1990): 351–58.

Sinha, R., R. Kumar, S. Sinha, S. K. Tandon, and M. R. Gibling. "Late Cenozoic Fluvial Successions in Northern and Western India: An Overview and Synthesis." *Quaternary Science Reviews* 26 (2007): 2801–22.

Sloan, L. C., and E. J. Barron. "Equable Climates during Earth's History." *Geology* 18 (1990): 489–92.

Smith, R. *Catastrophes and Disasters*. Edinburgh: W. & R. Chambers, 1992.

Smith, R. B., and L. W. Braile. "The Yellowstone Hotspot." *Journal of Volcanology and Geothermal Research* 61 (1994): 121–87.

Spignesi, S. J. *The Odd Index, the Ultimate Compendium of Bizarre and Unusual Facts*. New York: Plume Books, 1994.

Staubwasser, M., F. Sirocko, P. M. Grootes, and M. Segl. "Climate Change at the 4.2 ka BP Termination of the Indus Valley Civilization and Holocene South Asian Monsoon Variability." *Geophysical Research Letters* 30 (2003): 1425.

Staubwasser, M., and H. Weiss. "Holocene Climate and Cultural Evolution in Late Prehistoric-Early Historic West Asia." *Quaternary Research* 66 (2006): 372–87.

Steel, D. "Planetary Science: Tunguska at 100." *Nature* 453 (2008): 1157–59.

Stockton, C. W., and D. M. Meko. "Drought Recurrence in the Great Plains as Reconstructed from Long-Term Tree-Ring Records." *Journal of Climate and Applied Meteorology* 22 (1983): 17–29.

Strothers, R. B. "Mystery Cloud of AD 536." *Nature* 307 (1984): 344–45.

Tannehill, I. R. *Hurricanes (Their Nature and History, Particularly Those of the West Indies and the Southern Coasts of the United States)*. Princeton, NJ: Princeton University Press, 1944.

Taylor, J. *Petra*. London: Aurum Press, 2005.

"The Thames." *London Times*, February 2, 1814, p. 3.

Thomson, D. P. *Introduction to Meteorology*. Edinburgh and London: W. Blackwood and Sons, 1849.

Tishkoff, S. A., E. Dietzsch, W. Speed, A. J. Pakstis, J. R. Kidd, K. Cheung, B. BonneTamir, A. S. Santachiara-Benerecetti, P. Moral, M. Krings, S. Paabo, E. Watson, N. Risch, T. Jenkins, and K. K. Kidd. "Global Patterns of Linkage Disequilibrium at the CD4 Locus and Modern Human Origins." *Science* 271 (1996): 1380–87.

"Trent Tidal Bore." *London Times*, September 16, 1935, p. 9.

Turco, R. P., O. B. Toon, T. P. Ackerman, J. B. Pollack, and C. Sagan. "Nuclear Winters—Global Consequence of Multiple Nuclear Explosions." *Science* 222 (1983): 1283–92.

University of Chicago News Office. "Tetsuya 'Ted' Fujita, 1920–1998," press release, November 20, 1998, http://www-news.uchicago.edu/releases/98/981120.fujita.shtml (accessed July 28, 2007).

Vose, R. S., C. N. Williams Jr., T. C. Peterson, T. R. Karl, and D. R. Easterling. "An Evaluation of the Time of Observation Bias Adjustment in the U.S. Historical Climatology Network." *Geophysical Research Letters* 30 (2003): doi:10.1029/2003GL018111.

Walker, G. T., and E. W. Bliss. "World Weather V." *Memoirs of the Royal Meteorological Society* 4 (1932): 53–84.

Waller, D. *The Pundits: British Exploration of Tibet and Central Asia*. Lexington: University Press of Kentucky, 1988.

Ward, K., ed. *Great Disasters*. Pleasantville, NY: Reader's Digest Association, 1989.

Webster, N. *A Brief History of Epidemic and Pestilential Diseases*. 2 vols. New York: Burt Franklin, 1799; reprinted, 1970.

Weems, J. E. *The Tornado*. Garden City, NY: Doubleday, 1977.

Weiss, H., and R. S. Bradley. "Archaeology—What Drives Societal Collapse?" *Science* 291 (2001): 609–10.

Whiteley, D., ed. *Handbook of Rock Art Research*. Walnut Creek, CA: AltaMira Press, 2001.

Whitford, J. "Has the Pacific Railroad Changed the Climate of the Plains?" *Scientific American* 21 (1869): 214

Winograd, I. J., J. M. Landwehr, K. R. Ludwig, T. B. Coplen, and A. C. Riggs. "Duration and Structure of the Past Four Interglaciations." *Quaternary Research* 48 (1997): 141–54.

Wyrtki, K. "El Niño—The Dynamic Response of the Equatorial Pacific Ocean to Atmospheric Forcing." *Journal of Physical Oceanography* 5 (1975): 572–84.

———. "The Response of Sea Surface Topography to the 1976 El Niño." *Journal of Physical Oceanography* 9 (1979): 1223–31.

———. "Sea Level during the 1972 El Niño." *Journal of Physical Oceanography* 7 (1977): 779–87.

———. "Sea Level Fluctuations in the Pacific during the 1982–83 El Niño." *Geophysical Research Letters* 12 (1985): 125–28.

Wyrtki, K., E. Stroup, W. Patzert, R. Williams, and W. Quinn. "Predicting and Observing El Niño." *Science* 191 (1976): 343–46.

Yates, T. L., J. N. Mills, C. A. Parmenter, T. G. Ksiazek, R. R. Parmenter, J. R. Vande Castle, C. H. Calisher, S. T. Nichol, K. D. Abbott, J. C. Young, M. L. Morrison, B. J. Beaty, J. L. Dunnum, R. J. Baker, J. Salazar-Bravo, and C. J. Peters. "The Ecology and Evolutionary History of an Emergent Disease: Hantavirus Pulmonary Syndrome." *Bioscience* 52 (2002): 989–98.

Zhang, J., and T. J. Crowley. "Historical Climate Records in China and Reconstruction of Past Climates." *Journal of Climate* 2 (1989): 833–49.

Zielinski, G. A., P. A. Mayewski, L. D. Meeker, S. Whitlow, M. S. Twickler, M. Morrison, D. A. Meese, A. J. Gow, and R. B. Alley. "Record of Volcanism since 7000 B.C. from the GISP2 Greenland Ice Core and Implications for the Volcano-Climate System." *Science* 264 (1994): 948–52.

INDEX